Chemical Analysis in the Laboratory
A Basic Guide

Chemical Analysis in the Laboratory
A Basic Guide

I. Mueller-Harvey and R. M. Baker
The University of Reading, Reading, UK

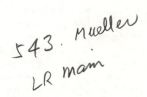

RS•C
ROYAL SOCIETY OF CHEMISTRY

ISBN 0-85404-646-1

A catalogue record for this book is available from the British Library

Published by The Royal Society of Chemistry,
Thomas Graham House, Science Park, Milton Road, Cambridge CB4 0WF, UK

Registered Charity Number 207890

For further information see our web site at www.rsc.org

Typeset by Computape (Pickering) Ltd, Pickering, North Yorkshire, UK
Printed and bound by Bookcraft Ltd, UK

Foreword

Analytical science is recognised as a key technology of critical importance to the needs of the chemical and pharmaceutical industries and of
sectors such as agriculture, food, medicine, environment, forensics,
engineering *etc*. It is truly an interdisciplinary science and maybe
considered as a virtual sector in its own right. Analytical science
provides the measurement information on which much government and
consumer legislation is based (*e.g.* levels of atmospheric pollutants from
motor vehicles and polyunsaturated fats in processed foods) and can be
considered as *the* problem-solving science. The analytical scientist has
to interact with scientists and technologists from both inside and
outside the natural sciences and also with the general public, all of
whom are reliant on the knowledge and skills of the analyst to help
solve their problems or to provide the relevant information. Any
analysis can be described in model terms by a combination of unit
processes, starting with a decision of the objectives for carrying out the
analysis and ending with an assessment of the validity of the data
produced and its relevance to the stated objectives. It is frequently the
primary role of the analytical scientist, in discussion with the client, to
ascertain the reasons for carrying out the analysis and the objectives
that are to be achieved. The breadth of knowledge of the anlytical
scientist will enable the correct route to be undertaken in order that the
objectives may be satisfied.

There are many good analytical textbooks now available, however
most concentrate on a detailed discussion of analytical techniques (*e.g.*
those based upon the principles of chromatography and spectroscopy),
and at the expense of the more fundamental considerations of why the
analysis is to be carried out and how the samples are to be taken. Whilst
most modern texts will introduce the reader to the importance of
sampling, many gloss over the serious errors which may be introduced
into the results if the sampling protocol is not undertaken in a logical
and statistically significant manner.

This book, written mainly from an applied biological sciences view-
point does attempt to address, in an interactive way, many of the

fundamental issues that affect analysis, particularly the initial decision making process used to define objectives and the assessment of the subsequent data, including how to estimate measurement uncertainty. The practicalities of sampling are well addressed, with examples quoted that relate to many popular agricultural and environmental situations. Other chapters consider the safety aspects of working in a laboratory, the care and use of normal laboratory equipment, sample preparation, manipulation of units and preparation of standard solutions. Some of these topics seem not to be taught much nowadays, but are vital to good results.

Although targeted at undergratuate students of the biological sciences (*e.g.* plant and agri-environmental sciences) who have to themselves make analytical measurements, parts of the book, particularly Chapters 1 and 2, will prove of immense value to all those new to analytical science, be they undergraduates, postgraduates or recent employees in an analytical laboratory.

Brian W. Woodget
UK Analytical Partnership, Skills Network Facilitator

Preface

Why is the Royal Society of Chemistry publishing a book covering basic information and exercises in chemical analysis? A few examples will serve to illustrate that there is indeed a need to address the very basic problems encountered in analytical chemistry laboratories through education and training not only in chemistry departments but also in other life science disciplines. Here are some excerpts from records in workfiles detailing customer requests:

- 'I have found this really unusual plant. Can you tell me what's in it for £20?' (a visiting scientist; an agronomist).
- 'Can you analyse nitrogens (*sic*) for me and let me know which tomatoes have been grown organically?' (biologist working in a food processing company).
- 'Just tell me the total P content in these bones . . . No, I don't want you to ash the samples . . . No, I can't give you the reasons for not ashing them as I don't want to prejudice you . . . You are the analyst, you should know what acid and what strength to use' (biologist working on bone research).

Given these real-life examples, there is clearly a need for improved training at the interface between scientific disciplines. Future users of chemical analysis need to be able to think 'analytically' and possess knowledge of basic analytical skills to help them when communicating with analysts. This involves the *whole* process of analysis: perceiving the idea, collecting the samples, performing the analysis, checking the validity of data, interpreting the data and reporting the results.

Analytical methods are useful research tools in the right hands. Students, researchers and other users of analytical methods need to appreciate:

- That each analytical technique has its advantages and disadvantages in terms of what information it can and cannot deliver (although this is one of the most basic principles of any investigative research or detective work, it needs emphasising nevertheless).

- That the wrong analysis request produces results that waste money (at best) and lead to wrong decisions (at worst).
- That there is no place for the outdated view that 'analysis is routine, easy and boring, *i.e.* not worth spending time or money on'. This was put simply and succinctly as 'rubbish in, rubbish out' (Gillespie *et al.*, 1999).

Analytical science requires a variety of skills and this book intends to address some of the basic issues by providing introductions, explanations and examples for teachers and lecturers to help students develop analytical skills. Reference is made to best practices and industry standards. Examples are drawn from the life sciences and aimed especially at students of the biological and environmental sciences. However, it is also intended for chemistry students, who, as future analysts, will be helping to solve problems presented by life science customers. Instrumental methods are not covered *per se* as these are covered in many other textbooks.

Before starting the analysis in the laboratory, the following basic issues need addressing:

- Adequate communication.
- Appropriate sampling, sample handling and preparation.
- Record keeping.
- Designing an experiment to answer questions within an available budget.
- Choosing an appropriate analytical method (*e.g.* recognising the difference between analysing total or available nutrients).
- Performing a risk assessment.

We have also found that one should not assume that all people working in laboratories have been taught the basics of chemical analysis:

- Good laboratory practices.
- Correct use of balances and volumetric glassware.
- How to prepare a standard solution and perform basic calculations (these are a common source of errors).
- When to use digestions or extractions of samples.
- The meaning of accuracy, precision and uncertainty of a measurement.
- How to assess if results are correct.

Our own and colleagues' experiences show that several areas warrant particular attention:

- Customer needs to be willing to communicate the objectives of the work to the analyst.
- Analyst needs to be able to communicate what information an analytical method can provide.
- Analyst needs to understand the researcher's or customer's objectives.
- Researcher and analyst need to ensure the correct use of blanks, standards, certified reference materials and understand the purpose of traceability, proficiency testing schemes and laboratory accreditation.

Finally, working at the interface between analytical chemistry and life sciences can be exciting and worthwhile. Just imagine the satisfaction felt by analysts and food scientists involved in research on leaves of *Moringa oleifera*, a drought resistant tree from the Sahel zone. These leaves have spectacular effects in reducing the malnutrition of children, improving their weight gain and reducing tiredness in adults (Fuglie, 2001). This interdisciplinary research – between analysts, nutritionists, clinicians and extension workers – was based on chemical analysis of vitamin A in fresh leaves, in traditionally prepared sauces (which had lost all vitamins) and in foods prepared with new recipes (which retained the vitamins). Thus the input of science improved the value of a traditional food and contributes to a sustainable use of local resources.

We hope that the material and examples presented in this book will stimulate students and teachers to look at their surroundings, develop their own analytical tools or exercises and thus engage in fruitful future dialogues between customers or users and the producers of chemical analysis.

References

Fuglie, L.J. (2001), *The Miracle Tree: The Multiple Attributes of Moringa*, CTA (Postbus 380, NL-6700 AJ Wageningen, The Netherlands) and CWS (West Africa Regional Office, Rue 8 Quart de Brie, Amitie 3, BP 5338, Dakar-Fann, Senegal), pp. 172.

Gillespie, A., Finnamore, J. and Upton, S. (1999), *VAM in the Environmental Sector*, VAM Bulletin 21, pp. 6–10.

Acknowledgements

The book is based on material originally prepared under 'Project Improve', funded by a grant from HEFCE, managed by the University of Hull. We have used many sources and apologies are due to those for whom we have overlooked specific acknowledgement. We are particularly grateful for the help, suggestions and material provided by:

Sarah Brocklehurst, College Analyst, Wye College, University of London (providing ideas and material for Chapters 1 and 3);

Professor Peter Keay, Director, Centre for Advanced Micro Analytic Systems (CAMAS), University of Luton;

Dr Richard Moyes, Director, Project Improve, School of Chemistry, University of Hull;

Dr Tina Overton, Assistant Director, Project Improve, School of Chemistry, University of Hull; and Project Improve workshops;

Dr Julian Park, Department of Agriculture, University of Reading (for information and exercises in Chapter 2);

Dr Geoff Potter, Department of Chemistry, University of the West of England (for information and material in Chapter 4);

Dr Elizabeth Prichard, LGC (Teddington) Ltd;

Dr David Rowell, Department of Soil Science, University of Reading (for use of material from Soil Science and Management course work in Chapter 5);

Steve Scott, Sartorius Ltd.

General points

We have used cm^3 throughout, instead of ml, for volume, and litre (l) instead of dm^3.

'Water' means deionised water of general laboratory standard, *i.e.* of conductivity less than $1\ \mu S\ cm^{-1}$, unless otherwise specified.

Chemical reagents are of 'specified laboratory reagent' quality, unless otherwise stated.

Contents

CHAPTER 1

Getting Organised for Useful Analytical Results

1 IMPORTANCE OF COMMUNICATION AND PLANNING

In the scenario described below, poor communication between an analyst and a researcher leads the analyst to do work on the samples that does not give what the researcher wanted and expected. Both the researcher and the analyst need essential information from each other to ensure that the work done on the samples is suitable, useful and well received by the researcher; and satisfying for the analyst.

While reading the scenario you are asked to consider the situation that the analyst has to address. At the end of it there are some explanatory notes but, before turning to them, you are asked to think about how the unsatisfactory outcome might have been avoided.

Suppose that the following conversation (which unfortunately is not entirely fanciful) took place between a researcher and an analyst in the laboratory:

Researcher: *I am going to harvest some plants from an experiment next week. I need to know the amount of P in them. Can you do that analysis for me?*

Analyst: *Yes, P determination in plant material is one of our standard procedures – when do you want to bring them in?*

Researcher: *Well, if I harvest them on Thursday, can I bring them first thing Friday morning?*

Analyst: *I'll check my work diary . . . yes that will be all right. That's agreed then, I'll see you on Friday morning, with the samples.*

However, there was a slight change of plan. The researcher had to leave early, but he entrusted the samples to a colleague to take them to the laboratory. So on Friday morning the samples duly arrived in five carrier bags with a note to the analyst saying:

1

'Here are the plant samples for analysis of P. I hope the results will be ready when I return in 2 weeks time'.

The carrier bags were labelled P0, P1, P2, P3 and P4, and each bag contained four young cotton plants, complete with roots and with soil adhering (Figure 1.1).

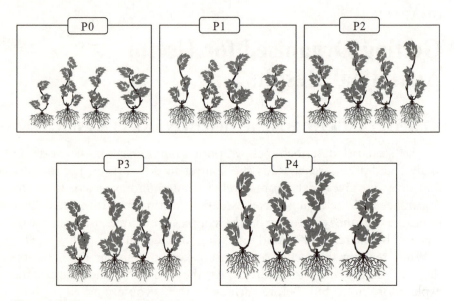

Figure 1.1 *Plants received for analysis*

Perhaps this analyst has not received samples from this researcher before and the researcher has not had an analysis done previously at this laboratory. So the analyst has to use her initiative in deciding how to deal with these samples.

The samples cannot be kept until the researcher returns:

- Do you foresee any difficulties for the analyst in making decisions on how to proceed?
- What would you advise her to do?
- What material should she take for analysis:
    - All the material provided?
    - Plants with soil removed?
    - Plants with soil and roots removed?
    - Leaves only?
- How many samples are there:
    - Five samples, one per treatment – each made by bulking together material from four plants?

- 20 samples – *i.e.* four replicates of one plant from each treatment?
- Do you think these decisions are important in relation to the purpose of the analysis or for any other reasons?

The work was at risk because the analyst and researcher each assumed too much of the other – at this stage they had not communicated adequately. Probably the researcher did not know enough about the process of analysis to question properly the service being offered. Probably the analyst is used to doing a particular procedure and assumed that it would match what was expected or needed in this case.

The analyst could not confirm any points with the researcher, but, being a sensible and helpful person, she proceeded on what seemed to her to be reasonable assumptions:

- Cut off the roots and soil from the plants.
- Take five aluminium trays, label them P0 to P4 and put the four plant tops from each bag into the appropriate tray.
- Put the trays in the oven to dry the samples at 80 °C overnight.
- Crush the dried plants in each tray and grind up all the material in a tray together to pass through a 1 mm sieve to make a homogeneous sample from each tray.
- Continue with the determination of the concentration of P (g kg^{-1}) in the five samples.

When the researcher returned, the laboratory was pleased to be able to report the results they had obtained (Table 1.1).

Table 1.1 *Analytical results*

Treatment	P0	P1	P2	P3	P4
P g kg^{-1}	0.96	1.12	1.12	1.26	6.35

Now did the researcher have some comments and questions:

- It looks as though the foliar spray treatment of P4 was very effective.
- Are the results statistically significant?
- There was about 1 kg of soil in each pot, so does the g kg^{-1} mean the same as g P taken up by the plant per pot?
- I want to work out what proportion of the P added was in the plants in each treatment.

What would you expect the analyst's responses to be?

- We didn't wash the plants so some of the P determined in the sprayed ones may be only dried P on the surface, but not absorbed in the plants.
  (Which probably means: you didn't tell me the plants were sprayed.)
- We can't do a test of statistical significance because there is no replication.
  (Which probably means: you should have labelled the plants as separate samples showing that there were four replicates of each treatment.)
- The g kg^{-1} means g P per 1000 g of dry plant material.
  (Which probably means: I thought anyone would know that.)
- We don't know how much dry material there was per pot so we can't calculate how much P was used per pot.
  (Which probably means: why didn't you weigh the plants or ask us to; or at least measure them in some way.)
- Here is our invoice for £50 + VAT.
  (Which probably means: well it's your loss, not mine.)

So both were very dissatisfied with the results of their efforts and no doubt each felt that it was all the fault of the other.

1.1 Activity

Take a few minutes to think about this situation. Is there any other information that the analyst should have and any other information that the researcher might think it is necessary to provide? List on a separate piece of paper as many points as you can that the analyst or the researcher need to know about the work requested:

- What the researcher needs to know.
- What the analyst needs to know.

1.2 Can We Avoid Misunderstandings?

Quite often a researcher has not thought through all that he or she is asking the laboratory to do and so is unprepared to provide all this relevant information. They may even feel that the analyst is being obstructive in demanding it. But you can see how things may go badly wrong if points are not clarified at an early stage.

Together the analyst and the researcher can work out a strategy to try to ensure that the analytical results obtained are as useful as possible to the purpose intended. It may be necessary for the researcher to organise transport from the field to the laboratory or to arrange for plot yields to be weighed in the field. Perhaps the samples will need to be kept in a refrigerator, pending their further preparation, so the analyst may need to organise that space is available in the refrigerator when required.

It would be wise of the researcher to write down the points of strategy agreed with the analyst and to leave a copy in the laboratory as confirmation of the arrangements. Such records form part of a workfile in a laboratory accredited to ISO/IEC 17025 (see Chapter 4, Section 6.1).

Questions:
- Why does the analyst want to know what types of plants are to be harvested?
- What problems may arise if the amount of sample material provided is (a) too large or (b) too small?

1.3 Notes on this Section

The researcher needs to know, for example:

- The set charge (if any) per sample for the test.
- Whether there may be a further charge depending on the amount of work required for sample preparation.
- An estimate of the time the analysis will take.
- The form in which the results will be presented.
- Whether, assuming the analysis goes well, the information produced will be useful and likely to add value to the experiment.
- If the amount of material in each sample is adequate or too little or too much.

The analyst needs to know:

- How many samples there will be.
- The deadline (if any) for reporting the results.
- What plants are being harvested.
- What material will be received (*i.e.* tops, leaves or whole plants).
- What will be the size of the sample (*i.e.* how may it be sub-sampled: fresh or dried).
- Whether the samples need washing (*i.e.* whether the plots will have received any sprays or dressings containing P).

- Whether any further work is required on the samples (*i.e.* are there implications regarding how the samples should be treated or stored?).
- Whether the researcher wants to know the P concentration in the plant material, or some measure of total plant content of P per plot (*i.e.* will the data collected allow the required calculations to be made?).
- If the weight of whole plants (fresh or dry) is needed what steps are to be taken to remove soil (or growing medium) and will the plants be really fresh when received in the laboratory?
- Overall, will the samples be taken in a satisfactory manner for the work to be done – especially with regard to organisation, labelling and avoidance of contamination?

The questions above raise the following issues:

- The analyst wants to know what plants are to be sampled because:
  - She may know or be able to look up the approximate range of contents to be expected.
  - She may know or expect some particular requirements in sampling, preparation or analytical method for certain types of plant material.
- Problems that may arise if the amount of sample material is (a) too large or (b) too small are:
  - Difficulty in dealing with a large bulk of samples in the laboratory due to limited size of bench space, containers and ovens available.
  - The small amount of sample taken for analysis may not be representative unless the whole sample is homogenised, which may be time consuming for large samples.
  - Amount of material may be insufficient for all analysis and checks required.
  - Small samples may be easily contaminated in the mill or grinder.

2 ONCE THE SAMPLES ARE IN THE LABORATORY

The researcher should send or bring the samples, with a sample list, having checked that the samples delivered match the list. Any discrepancies or missing samples should have been noted on the list, together with any special instructions agreed – *e.g.* freezing, cleaning, drying, *etc.*

On receipt, the analyst should first check the samples against the list, note any discrepancies, and check that any special instructions are taken in hand.

Suppose you are both the researcher and the analyst. You are not likely to have that initial conversation about the samples with yourself. However, you will still need to think through what you want to achieve from the analysis and how you must deal with the samples to achieve it. You must ensure that the points such as you have listed are addressed, and that the relevant information is being taken into account.

2.1 Keeping Records (DON'T FORGET – that you will not remember)

For useful tips, see Rafferty (1999).

- Have a book, preferably hard backed, in which to write all ideas, plans, discussions with your supervisor and laboratory staff, advice, instructions, problems encountered, sample lists and experimental results.
- Put your name on the outside of the book and your name and address on the inside.
- Make sure that you always date your work.
- Do not keep records on scraps of paper – they will get lost, mislaid, wiped clean in the laundry or used to mop up spills.

3 PROBLEMS WITH SAMPLE MATERIAL

Usually we are concerned to get a sample that is 'representative' of the whole of a bulk material, or a fluid or a tissue. Sampling has to take account of the fact that only a small amount of sample is collected from a relatively vast bulk. Often it is necessary to take a number of random samples throughout the material. Any portion of the material represented by the sample must have an equal chance of being included. If we sample only from the parts that are easy to reach, we may be introducing bias.

On the other hand, we may want to introduce some selective sampling, for example, to avoid some non-representative areas – such as the gateway or headlands in a field. Again, we may want to divide the bulk material into different sampling units – such as from different areas of a warehouse or at different heights in the hold of a ship or in a silo or at different depths in a soil.

In the laboratory, we are more often concerned with taking a test sample that is representative of the laboratory sample received. We need first to create a homogeneous laboratory sample from the material submitted to reduce the variability between different test samples. Depending on the type of material involved, sample preparation, sample storage and sub-sampling may all have some influence on the analytical results eventually obtained. It is essential that the researcher and the analyst both understand exactly what is wanted from the analysis in order to avoid inappropriate and irretrievable decisions at this stage.

Some samples do not remain well mixed when dried and handled in the laboratory. For example, a sample composed of sand and compost will easily separate out into its different components simply on standing for a few weeks, so it needs to be thoroughly mixed each time a test sample is taken. Samples that contain a mixture of coarse and fine particles are similarly difficult to keep homogeneous. A sample made up of whole plants will, when dried, become a heterogeneous mixture of leaves and stems. It requires special care to keep the sample together and mill it down to make a uniform material.

The bulk of a sample received in the laboratory may be too big for convenience of storage or for further preparation. We have to decide how and at what point the sample size can be reduced or sub-sampled. If the entire sample received can be homogenised then subsequent sub-sampling is not a problem. Otherwise special care is required to ensure that the sub-sample taken is again fully representative of the material received.

With very wet samples – such as silage or sediments – there is the potential for the soluble components of interest to be lost during sample preparation. Some components *e.g.* ammonia or organic compounds may also be volatilised or metabolised during storage and preparation. Such samples may need to be frozen for storage and kept frozen throughout the sample preparation and as test samples are taken.

Microbiological activity may continue in the samples and change the constituents, *e.g.* changes in nitrate content of soil and extracts unless they are stored at below 4 °C. Samples may be altered if they are dried too severely. Silage material, for example, may show a reduction in N content if it is dried at 100 °C.

By convention, soil analysis is done on 'fine earth'. This is air-dried soil, ground and sieved through a 2 mm screen after removal of stones and roots larger than 2 mm diameter. Unfortunately there is not always an easy distinction between 'stones', 'weathered rock' such as soft chalk and hard lumps of soil. So it may be difficult to standardise the preparation of the laboratory sample.

4 IS THE AMOUNT OF WORK REQUIRED FEASIBLE?

Next we will look at the question of what we can do if the work required would take too long or be beyond the physical or financial resources available.

Suppose you want to determine the changes in nutrient contents of a grass crop from field trials at six locations – each with factorial treatments of two levels (0 and 1) of N, P and K fertilisers. *i.e.* eight treatments:

$$\text{N0K0P0} \quad \text{N0K1P0} \quad \text{N0K0P1} \quad \text{N0K1P1}$$
$$\text{N1K0P0} \quad \text{N1K1P0} \quad \text{N1K0P1} \quad \text{N1K1P1}$$

For the agronomic management aspect of the trial the plots would be cut immediately before the fertiliser treatments are applied and then another 12 times at two-weekly intervals. The agronomists would dry and weigh the whole cut from each plot and then hand the material over for the analytical studies.

For the investigation of plant nutrients, the draft plan is to take from each cut a sample for analysis of N, P, K, Ca, Mg, B, Fe, Mn, Cu and Zn. So there would be over 100 samples per month for six months – April, May, June, July, August and September. You would need to check with the laboratory to see if they can handle that many samples over that period.

To do the full study as proposed, there would be: 8 treatments × 6 locations × 13 sample times × 10 determinations = 6240 determinations.

Assuming that the laboratory can deal with the samples, accepting that their charges are: £2 per determination plus £1 per sample for sample preparation; then this will cost £13104.

But suppose you have only £1200 available in your project for the analysis – decide how you would spend that amount to get as much useful information as you can from the trial. You will have to re-appraise the requirement for analysis with a view to cutting down the number of samples or determinations – perhaps by reducing the number of sampling times or bulking some samples together.

- You can save £1 per sample if you can do all the sample preparation work for the laboratory.
- What precautions have to be taken in bulking material from different cuts?
- Will you be able to calculate the total removal of nutrients in the grass cut over the period of the experiment?

Think about the questions raised and the possible options you have before looking at the proposed strategy in the notes below.

4.1 Notes on this Section

Various possible options may lead to alternative solutions but a suggested strategy to work within the resources available is as follows:

- You agree to do all the sample preparation work – saving £1 per sample.
- We have to cut down on the number of samples. While we want to keep sufficient replication, the main objective is to investigate the effects of applied fertilisers on nutrient uptake, rather than a comparison of locations – we can choose to do analysis on three locations, rather than six.
- If we bulk together too many cuts we may miss or dilute any treatment effect, however it is reasonable to analyse the initial cut (before fertilisers) and then to combine the material from each following two consecutive cuts. This will give a total of seven sampling times rather than thirteen. So now we have: 8 treatments $\times$ 3 locations $\times$ 7 sample times = 168 samples.
- Suppose we do only N, P and K analysis on all samples, this will amount to 504 determinations; then we could do the remaining analyses on the control plots (N0P0K0) and the complete fertiliser plots (N1P1K1) and only on the initial cut and a composite of the following 12 cuts. This would give an additional: 2 treatments $\times$ 3 locations $\times$ 2 sample times $\times$ 7 elements = 84 determinations.
- This would give a total of 588 determinations, costing £1176.

Of course there is a loss of information in the study of secondary and micronutrients (Ca, Mg, B, Fe, Mn, Cu and Zn). On the other hand the study would still be able to show whether the soils can supply these nutrients adequately to support any increased yield of grass due to the fertilisers or else that the grass becomes depleted in these nutrients.

With regard to the bulking of material from different cuts, each cut is likely to be too bulky to be kept whole. Therefore it would need to be thoroughly homogenised and then a convenient-sized sub-sample taken from it. When bulking two or more samples together, this should be in proportion to the yields of the plots, to preserve the ability to calculate the total nutrient removal in the grass over the period of the experiment.

5 PROBLEMS WITH DEFINITIONS

Misunderstandings between the analyst and the researcher (or student or client) can easily remain undetected unless both are willing to talk to each other. The analyst can expect that a non-analyst will have some misconceptions about analytical work. Though exactly what the misconceptions are may be quite unpredictable. So the onus is on the analyst to find out as closely as possible what the student, client or researcher really wants.

It is important to realise that some words commonly used in analytical chemistry may convey a different meaning to non-specialists or specialists from other disciplines or when used outside the context of analysis. Indeed, it may surprise some non-chemists to learn just how imprecisely defined some terms are that we use in an apparently scientific context. We will look at some examples in more detail.

5.1 Empirical Methods

If we are asked to determine the amount of nickel in a sample of a metal alloy, we would expect that nickel is a specific component of the alloy and that the absolute amount of it could be measured. In many fields of analysis the names of some parameters that we measure refer only to a part of a component entity in the sample. Such a name does not precisely identify a specific part of the sample, but is related to the particular method used. These are empirical methods – they have been arrived at by a large number of experiments to provide results that are useful in certain practical applications. The results can be interpreted to formulate diagnoses, treatments or management decisions. The justification and basis for such methods is that they can give results that have been found to correlate with some important performance characteristics. Unfortunately though, they seldom increase our understanding of the mechanisms of action or processes involved.

5.1.1 Available Nutrients. For example the terms 'available' or 'plant available' P are often used in relation to soil testing. It is usually taken to mean the amount of P that the soil can supply to the crop in a growing season, but it is not necessarily defined in relation to where, and in what form, the P is in the soil.

To make useful determinations of the fertility of soils or of the fertiliser requirements for crops, a laboratory must use test methods that are suitable for the specific types of soils it receives. It will have been shown that the test results correlate reasonably well with the ability of these soils (perhaps chalky soils in England) to supply P to the growing crop.

Unfortunately the particular method that the laboratory normally uses to determine 'available' P may be inappropriate for different types of soil being studied in a research project (which might include tropical ultisols or oxisols).

Because we are aiming to extract available nutrients, the amount extracted should correlate well with the amount taken up by the plant. There are several different extractants recognised for determining 'available' P in the soil (Table 1.2). Those listed are well known and field correlation studies have been done in some areas of the world. However, these methods have been used widely and interpretations are attempted in a great range of conditions and under various land husbandry systems, frequently without any confirmation that the assumed correlations apply in practice for the particular situation.

Table 1.2 *Extractants for 'available' P*

Name	Extractant – dilute solutions of	Intended for
Bray 1	$NH_4F + HCl$	non-calcareous soils – often used for tropical soils
Mehlich 1	$HCl + H_2SO_4$	acid and sandy soils
Olsen	$NaHCO_3$, pH 8.5	neutral and alkaline soils, but used widely

An experimenter may be studying soils and crops that are quite different from those that the laboratory normally receives. In this case the usual laboratory method may give a very poor indication of the 'availability' of P in the soil.

5.1.2 Fibre and Lignin. In ruminant feeds, such as hay, grass and silage, a large proportion of the material comes from the cell wall tissues which are composed mainly of cellulose. In the assessment of the nutritional value of the feed materials by analytical methods, it has proved useful to measure total cell wall contents. Analytical methods have been devised to determine crude fibre, neutral detergent fibre, acid detergent fibre and lignin.

These named fractions do not refer to well defined chemical constituents or anatomical components of the plant material. Their compositions vary from one type of plant material to another. They are identified entirely by the method of determination, so the methods of analysis used must be carefully specified and followed in detail for results to be consistent and comparable from one batch to another.

5.2 Generic Terms

There are some generic terms (like carbohydrates, proteins, volatile fatty acids, PCBs) that are used to describe a group of compounds or components. The analyst needs to be sure whether it is the total amount of the generic group that is to be measured in a sample or rather some specific compounds that are of importance for the current study.

5.2.1 Tannin. Tannins, for example, are a large group of phenolic compounds that occur in many plants and trees and can be part of the diets of browsing animals. It is well known that tannins may influence the digestibility of protein and fibre in the animals. Analysts are often asked to measure 'tannins' in animal research projects concerned with the dietary quality of browse material. The laboratory would usually then measure the total contents of tannins in the materials being studied. However, in comparing different types of browse materials, it may be more important to know their contents of particular tannin compounds – of which some can increase nitrogen uptake, whereas others markedly decrease it.

5.2.2 Clay. The clay content of soils is defined as the fraction of soil particles that are less than 2 μm in diameter. Because small particles have a relatively large surface area, it is in the clay fraction, and especially in the finest part of the clay and associated organic matter, that most of the chemical and physical properties relating to soil fertility reside. Measurement of the total content of clay in a soil is frequently useful in understanding or predicting the behaviour of soils in response to fertilisers, tillage, irrigation, *etc.* Sometimes, however, identification of the particular types of clay minerals dominant in the soil may be required because the different clay minerals give very different properties to the soil as a whole.

As an analytical task, identification of the clay minerals is quite different from determining the total content of clay.

5.2.3 Nitrogen. Most of the nitrogen in plant material, animal feeds and soil is in the form of protein – so much so that the common method of determining protein in foods and feeds is to determine total N and apply a multiplication factor (*e.g.* 6.25) to express the result as protein. However, most of these materials do contain some inorganic nitrogen as ammonium compounds, nitrates or nitrites. The analyst must know whether or not the inorganic components are to be determined separately.

Question:

In a genuine enquiry a laboratory was asked to determine organic and inorganic nitrogen in tomatoes to prove whether or not they were organically produced.

How would you expect the laboratory to respond to such a request? Just give yourself a moment or two to think about this before reading the explanation below.

Explanation:

Fortunately the customer did tell the analyst what he expected the results to tell him, otherwise the laboratory might just have gone ahead with the determinations of organic nitrogen (as crude protein) and inorganic nitrogen (as ammonium, nitrate and nitrite).

The customer was demonstrating the confusion of a non-chemist encountering the familiar words 'organic' and 'inorganic' but not appreciating the different contexts in which they are being used. On the one hand, 'organic' or 'inorganic' may refer to the type of fertiliser materials applied to the crop; on the other hand it may refer to the chemical composition of the crop itself.

The analyst had to explain to the customer that the crop can absorb nitrogen only in inorganic form (nitrate and ammonium). So organic nitrogen applied to the soil as manure or compost has to be converted into inorganic form (*i.e.* mineralised) in the soil before the crop can use it. Within the crop, most of the absorbed inorganic nitrogen, from whatever source, is converted through the natural biochemical reactions into organic nitrogen (*i.e.* protein).

So the determination of different forms of nitrogen in the crop cannot prove or disprove that the crop was 'organically' grown.

CHAPTER 2

The Sampling Plan, Sample Collection and Preparation

1 THE SAMPLING PLAN

Anyone taking samples for analysis needs to realise the importance of stating adequately the problem they want to solve or identifying the basic questions that they want the analysis to address. Then they need to assess whether their plan is capable of giving results that will provide the answers. We should ask ourselves:

- What should we sample and what do we need to measure?
- How many samples would there be for chemical analysis?
- Can we ensure that results obtained can be analysed statistically?
- What are the limitations imposed by considerations of cost, convenience, accessibility and equipment?

We cannot analyse all the material we are studying. As we have discussed in Chapter 1, we usually want to get a sample that is 'representative' of a bulk material or a fluid or a tissue being studied. The practical solution is to take a number of samples throughout the material. For sampling to be truly random, any portion must have the same chance of being included. We need to avoid introducing bias; if we are taking samples from animals, we should not pick only the ones that don't run away; if we are sampling a forest, we need to avoid concentrating on areas near the roads. Random sampling allows us to use some statistical techniques on our results and to estimate the range about our sample mean within which the 'true' mean probably lies.

You may be taking samples which you will pass on to an analytical laboratory for analysis or you may be going to analyse them yourself. In either case there has to be a clear continuity of purpose from the sampling, through the analytical stages to the final report and conclusions. Bear in mind that the analytical result may depend on the method used but it certainly depends on the sampling plan. The essential questions are:

- Where to collect the samples?
- When to collect them?
- How many samples to collect?

The design of the sampling plan must also take into account whether the system to be sampled is static or dynamic. If we are sampling pollutants in a river, the concentration may change over periods of a few minutes or over several days. In such a case we might take small samples regularly and composite them for each day. We would probably need to include all periods of time including weekends and nights. However, it may be the variation throughout the day, the week or the month, that we particularly want to investigate. In studying traffic pollution, for example, or pollen counts, it may be useful to know the peak periods throughout the day at different locations.

We want to identify the goal of the study to make the sampling plan appropriate to the purpose. Measurement procedures may influence the plan, so these need to be decided in advance. Sampling must be done with the requirements of the analytical method, and the evaluation of the results, in mind. Organic materials can be altered by bacteria – so a preservative may be needed or samples may need to be frozen. Sample size must be adequate for all the work required.

Monitoring the environmental safety of a large rubbish dump or a landfill site might be concerned with toxic effluent, heavy metal content or methane production. Limits can be imposed on the amounts of hazardous materials that the site may hold. Usually such sites are very heterogeneous and it is hard to tell how many samples would be required to determine an overall average value for a particular parameter. We might need a pilot study first to determine the intensity and distribution of a pollutant we are investigating. However, if the sampling plan allows us to evaluate the results statistically, then we can put confidence limits on our estimate of the average value (see Chapter 5, Section 6).

1.1 A Landfill Example

Suppose that a landfill site contains PCBs (polychlorinated biphenyls) as a pollutant and we have to reduce the average concentration of them over the site to below 1%. The effective treatment for PCB contamination is to remove the soil, incinerate it and return it to the site.

Because of the irregular way that waste materials are added to a landfill, we would expect the PCBs to occur in intense irregular patches. This perception can be expressed as a hypothesis that the site is contaminated but that much of the site has low or nil concentration of PCBs and does not require treatment. The corollary is that some parts of the site are of highest priority for treatment. We need, then, to develop a rational strategy for sampling and analysis that gives a cost efficient indication of which areas need treatment and which do not. As the distribution of contamination is very heterogeneous, the sampling plan must provide an estimate of the probable range (with confidence limits) within which the 'true' average falls. When this range is below the action threshold level of 1%, no further treatment is required.

For example, the landfill site could be divided into 0.25 hectare target areas (*i.e.* 50 m × 50 m). Then a target area might be divided into a grid of 25 sampling cells of 10 m × 10 m (Figure 2.1).

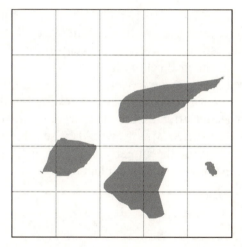

Figure 2.1 *Target area of landfill site showing patches of pollution and sampling grid*

Within each grid cell we could take three random samples and calculate an overall average for the target area (Figure 2.2) from the analysis results for the 75 samples. This stratified random sampling design allows statistical analysis of the results. We can calculate an upper confidence limit of the overall average concentration and test the 'null' hypothesis that the soil is contaminated above the action threshold of 1%.

If the average concentration, at the 95% higher confidence limit, is below the action threshold for treatment, the target area can be

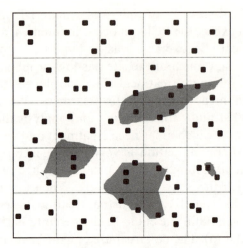

Figure 2.2 *Landfill target area with three random sampling positions in each grid area*

designated as 'clean'. That is to say that the 'null' hypothesis can be rejected and no treatment is required. Otherwise, if the higher confidence limit of the estimated average concentration of PCBs is above the action threshold, the grid cell with the highest concentration would be selected for treatment. A new overall average and confidence limit would be recalculated with the test sample results for the treated unit set to the level expected after treatment (perhaps zero concentration). The next most contaminated squares would be selected and the overall average recalculated, until the upper confidence limit falls below the required action threshold (in this example, below 1%).

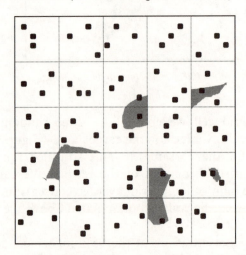

Figure 2.3 *Target area of landfill site after treatment of selected grid areas*

When the selected squares have been treated, the pollution in the target area might be as shown in Figure 2.3. If the remaining estimated average overall concentration of PCBs is below the action threshold at the 95% confidence limit, then no further treatment is needed.

Greater intensity of sampling, by using a finer grid or by taking more samples per sampling unit, would give a better prediction of which areas require to be treated, especially if the pollutants were more strongly localised.

On a real site, however, we would not know how the pollutant was dispersed and we would not be able to develop a customised sampling strategy. Instead, we would have to make some estimates or assumptions about the general character of the pollution and develop a general design (as above) that would apply to each target area of the site.

If the combined cost of sampling and analysis is relatively low compared with the cost of treatment, then more intensive sampling would save on treatment costs, because less of the landfill area would be treated unnecessarily. Of course the converse applies too – if the cost of treatment is very low compared with the cost of analysis, then intensive sampling might not be cost effective. However, there might be other reasons for wishing to reduce the treatment as much as possible. For example, there may be some environmental impact associated with the treatment itself.

A range of sampling strategies for this type of situation is the subject of a simulation exercise on the internet which can be found at: *http://etd.pnl.gov:2080/DQO/simsite/home.htm.*

2 EXERCISES

Devise a sampling plan for one of the following investigations or experiments. In each case, the investigation can be stated in terms of a hypothesis that is to be tested. This helps to define the problems and objectives clearly. Ask yourself the types of questions put in the opening paragraph of this chapter.

2.1 Soil Nutrients and Tree Seedlings

Nutrients contained in the seeds of trees may be entirely sufficient for their germination and the early growth of seedlings. Design an experiment to test the hypothesis that the nutrient content of the growing medium (John Innes potting compost) will not affect:

- The germination of seeds.

- The growth of seedlings during the first month of growth.
- The nutrient contents and nutrient uptake in the aerial parts of seedlings during the first month of growth.

Information:

- John Innes composts are a mixture of sterilised loam soil, peat and sand containing various amounts of nutrients and chalk. The nutrients are usually added in the form of a pelleted NPK fertiliser (Osmocote).
- John Innes seed compost contains no added nutrients.
- John Innes No.1 potting compost contains fertiliser at 3 g per litre and chalk at 0.5 g per litre.
- John Innes No.2 potting compost contains fertiliser at 6 g per litre and chalk at 1.0 g per litre.
- John Innes No.3 potting compost contains fertiliser at 9 g per litre and chalk at 1.5 g per litre.

2.2 Heavy Metal Contamination of Farmland

You are asked to study the distribution of zinc and lead in farmland soil and vegetation adjacent to a motorway and to test the hypothesis that the motorway traffic is (or has been) a source of contamination by these metals.

- With regard to the statistical analysis of the results, what will be your 'null' hypothesis?
- Will the concentration of the metals vary with depth in the soil?
- Will the concentration of the metals vary with the distance from the motorway?

2.3 River Pollution

You need to investigate the total 'load' of the inorganic contaminants nitrate, ammonium and phosphate carried by a river, perhaps to estimate the amounts coming from farmland or to study the impact of a sewage works (wastewater treatment plant). There are public health limits for ammonia and nitrate (and nitrite) in drinking water. If the concentration is high in raw water, the cost of treatment is increased. All of these nutrients cause eutrophication of lakes and backwaters. Ammonia is toxic to aquatic life – depending on the concentration, pH, temperature and the organisms present.

- A sewage works would be a 'point source' of pollution, *e.g.* from an effluent pipe; whereas agricultural pollution is likely to be from a non-point source.
- You will need to find the concentration of dissolved materials (expressed as grams per cubic metre of water) and the flow rate (expressed as cubic metres of water flowing past a point along the river each second). The 'load' is the number of grams of a dissolved nutrient flowing down the river each second (*i.e.* the flow rate multiplied by the concentration). But the flow rate is not constant across the width or throughout the depth of the river and the concentrations of dissolved materials may not be constant either.
- What would you do about particles in suspension which may contain the nutrients too?
- You might want to study seasonal changes or to monitor how much of these nutrients are being carried to the sea.
- Sedimentation may cause contaminants to accumulate in slow moving areas.
- Concentrations may vary downstream through further additions or losses of pollutants (through sedimentation and biological processes).

2.4 Notes on the Exercise

2.4.1 Soil Nutrients and Tree Seedlings. In the germination experiment you could, for example:

- Compare composts having at least two levels of nutrients (N0 and N1).
- Use petri dishes or beakers containing at least 10 seeds each.
- With at least two replicates.
- Moisten the compost, cover the dishes and measure the percentage germination after two weeks.

In the growth and nutrient uptake experiment, the options might be as follows:

- Germinated seeds from the seed compost (no added nutrients) would be transplanted to a similar arrangement of replicated pots, containing at least two levels of nutrients and kept watered with deionised water. You would need to decide how many seedlings there should be per pot and how you will sample the seedlings. You could choose to sample only after two months; or after one month and two months of growth.

- To determine nutrient uptake (above ground parts only) you would need to weigh the cut seedlings and determine the dry matter content.

2.4.2 Heavy Metal Contamination of Farmland. The 'null' hypothesis is that the motorway has no effect on metal contents of the soil and vegetation. The null hypothesis can be rejected if zinc and lead contents vary systematically with distance from the motorway. Samples of vegetation and soil should be taken on a transect perpendicular to, and on both sides of, the motorway. For example soil samples could be taken from the surface and at 20 cm depth at each 10 metres from the perimeter fence adjacent to the motorway to a distance of 100 meters from the fence.

2.4.3 River Pollution. The sampling plan has to allow for the variations in flow rate and concentrations of nutrients in different parts of the river. So it is not adequate simply to sample the middle of the river. The study might be related to the effects of sewage works (anticipating increased loads of ammonia and phosphate or the suspected leaching of nitrate from agricultural land).

Confluence of tributaries might reduce concentrations, but still increase the load. You might want to investigate the variation with season and with time of day. Perhaps it is the amount of contaminants that are being carried to the sea that is important – or the build-up of contaminated sediment at certain locations.

3 TAKING SAMPLES

It is important to be prepared and properly equipped for the task in hand. The sampling plan usually provides a logical basis for identifying and labelling the individual samples. Labels must be water stable and be well attached to the sample containers. Sampling tools will vary according to the type of material and the investigation. Scissors, trowels, spades and augers, corers, ladles, scoops, syringes, scalpels, buckets, jars and vials, among others, may be used at different times. Containers used for holding the samples are similarly varied although polythene bags, cartons and plastic bottles are frequently both suitable and convenient.

Samples must not come into contact with any material that may contaminate them or otherwise invalidate the results subsequently determined. Such potentially contaminating material includes other samples, so the tools and containers must be cleaned each time they are used. Once the samples have been taken they must be kept in conditions that prevent their deterioration and protect them from all possible sources of contamination. Sieves, trays, oven shelves, benches *etc.*, are all suspect. Trays and shelves, for example, may be made of galvanised steel which could invalidate subsequent determinations of zinc in the samples.

3.1 Composite Samples

Frequently a sample that represents a bulk material or an area or experimental plot is made up of a number of 'increments'. This is the usual way of sampling soil from a field (see Figure 2.4). In the landfill example above, each random sample could be a composite of several increments around the sampling point. This procedure helps to prevent small-scale variability in the sampled material from having unwanted influence on the results.

By taking several increments, the total amount of material accumulated may be too much for a single sample. The bulk has to be reduced. For solid materials this is done by mixing the whole sample together thoroughly, taking a conveniently sized portion from it and discarding the rest.

3.2 Handling and Storage

The samples must be stored in good conditions pending analysis. The storage conditions that are appropriate will depend upon the type of sample and the determinations required. Most soil analysis, for example, is done on air-dried soil, but the nitrate content can be altered by drying, so nitrate studies are usually done on fresh soil. If the soil has to be stored, this should be at <4 °C and for no more than 48 hours. Similarly samples in which analysis of the volatile components is required, such as silage, urine and faeces, are usually kept frozen until immediately before the analysis.

Many samples that we encounter are perishable or otherwise subject to natural deterioration, and so we have to ensure that our sample handling procedures are appropriate for the analytical work required. The sampling plan must take account of any determinations that are particularly susceptible to the handling and storage conditions of the samples. Special arrangements may be needed for the rapid transfer of samples from the field to safe conditions in the laboratory.

3.3 Sample Preparation

In the laboratory, samples should be given a unique identifying number and recorded in a sample register. Then samples from different sources should not become confused with each other. The laboratory number may consist of a batch number and a sequence number. By using the laboratory number, there is no need to copy all the sample details each time the sample is taken for another test.

Sample preparation may involve filtration or centrifugation of liquid samples or drying, grinding and sieving of solid samples. Some sorting of material in a sample may be done in the laboratory – for example it is usual for stones, large roots and undecomposed vegetation to be removed from soil samples. Clearly, sample preparation must be done in full knowledge of what the subsequent work requires. It is no good looking for the heavy metals carried by a sample of river water if they were adsorbed onto the sediment and have been discarded with it.

Oven drying is a common procedure in the preparation of vegetable samples, but we need first to check whether fresh weights or original moisture contents will be required for the subsequent evaluation or reporting of the results.

When we take a test sample from the laboratory sample, again we want the test sample to be representative of the whole of the laboratory sample. Components of different densities or different particle sizes in liquid or solid samples become segregated in storage. Usually we need to mix the laboratory sample thoroughly, to ensure that it is homogeneous, each time we take a test sample from it.

3.4 Contamination

At all stages we need to be aware of the possibility of contaminating the sample and precautions necessary to avoid it. There is a high risk of contamination during sample preparation in the laboratory. Grinders made of stainless or hardened steel, for example, may increase the contents of iron, manganese, nickel and chromium in the sample. Other grinding materials may give other contaminants. This can be a problem in some trace analysis and it may be necessary to be especially careful in the choice of grinder and the intensity of grinding.

Some polythene and polypropylene containers may give iron, manganese, calcium, copper and zinc contamination to abrasive samples or to solutions kept in them. For the most rigorous trace analysis, extra precautions and special apparatus may be required (for comprehensive information on this topic see Prichard *et al.* (1996)).

4 PRACTICAL EXERCISES

One of the following investigations will provide the samples for practical analytical work described later.

4.1 Tea Sampling

It is possible to buy tea whose region of origin is clearly shown on its label *e.g.* Kenyan, Darjeeling, Assam.

Make a sample plan for an investigation to find out whether or not:

- Teas that are from apparently different sources contain different amounts of zinc and manganese.
- Cups of tea made from them contain different amounts of these elements.
- Finely grinding the tea to a uniform particle size is worthwhile – in order to reduce the variability between duplicate determinations.
- The finely ground tea releases more of the elements into the brew.

The plan for the investigation should allow for no more than 100 determinations.

Why is it preferable to reduce the variability between duplicate results as much as possible?

4.2 Soil Sampling

Do an investigation into the variability of soil P between different locations, soil types or different parts of a field. The ultimate objective might be to adapt fertiliser applications to specific sites or to select sites for further experimentation.

Aim: to take soil samples representing three types of cultivated land (*e.g.* three different fields, different soil types or three positions on a slope) at depths of 0–20 cm and 20–40 cm.

Each sample will be made up of soil at a single depth (0–20 cm or 20–40 cm) from up to 10 individual sampling spots within one sampling area. Each sampling area will represent a single replicate of one type of land. You will need at least two replicates, so there must be at least two

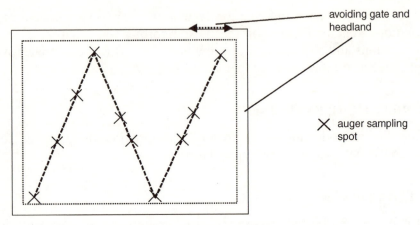

Figure 2.4 *Taking a soil sample from a field*

sampling areas for each type of land. For example to compare the P content of soil at the top, middle and bottom of a slope, one replicate might consist of sampling areas on the west side of the field and another replicate of sampling areas on the east side. Do you think there would be any advantage in taking replicate samples within each sampling area at each depth?

Equipment required: soil augers (screw type), trowels, buckets, polythene bags, labels.

Procedure:

- Select the areas to be sampled.
- Walk a 'zigzag' path across an area (Figure 2.4), taking 10 samples at intervals and bulk them together. A single bulk sample should represent no more than 1 hectare. For most purposes, gateways, headlands and other abnormal areas are avoided in making the sample representative, although sometimes additional sampling may be desirable to investigate these areas.
- At each auger sampling position screw in the auger to the initial depth (20 cm), withdraw the auger and remove the soil into a clean bucket.
- Clean the auger well, insert it carefully into the same hole and screw it down to the second depth (40 cm). Avoid contamination between successive vertical samples.
- Place the soil from the second depth into another bucket or bag. Experience shows that there is less variability in the soil at depth, so only about half of the sampling positions need to be sampled at the lower depth.

- For each area, and each depth in turn, mix together the accumulated soil and carefully sub-sample it down to a convenient amount. Keep the samples from each depth and each area separate, and clearly labelled.
- Make a sample list – for example, see Table 2.1.

Table 2.1 *Sample list for replicates I and II*

Sample number	Sample identity	Location	Depth (cm)	Replicate
1	S1D1$_I$	Site 1	0–20	I
2	S1D2$_I$	Site 1	20–40	I
3	S2D1$_I$	Site 2	0–20	I
4	S2D2$_I$	Site 2	20–40	I
5	S3D1$_I$	Site 3	0–20	I
6	S3D2$_I$	Site 3	20–40	I
7	S1D1$_{II}$	Site 1	0–20	II
8	S1D2$_{II}$	Site 1	20–40	II
9	S2D1$_{II}$	Site 2	0–20	II
10	S2D2$_{II}$	Site 2	20–40	II
11	S3D1$_{II}$	Site 3	0–20	II
12	S3D2$_{II}$	Site 3	20–40	II

4.3 In the Laboratory

Samples may need to be reduced in size, *e.g.* by crushing, screening and quartering. The final laboratory sample should be made homogeneous – by grinding, dividing and mixing. This is divided into sub-samples and ultimately test samples – the portions weighed out for the determinations required.

4.3.1 Tea Samples.

- Spread the samples onto aluminium trays.
- Dry overnight at 80 °C.
- Grind to pass through a 1 mm sieve.
- Store in labelled, sealed polythene bags.

4.3.2 Soil Samples.

- Spread the samples onto aluminium trays.
- Air-dry at room temperature (2–5 days).
- Crush by hand and remove stones and root material.
- Grind by hand in a pestle and mortar to pass through a 2 mm sieve and discard remaining roots and stones. (If the stones constitute more than about 10% of the sample then estimate the proportion of stones in the original sample.)
- Store the samples in labelled, sealed polythene bags.

4.4 Notes on the Practical Exercises

4.4.1 Tea Sampling.   The fine dust of tea leaves frequently shows higher contents of metal elements than the larger leaf material. The uneven distribution of the metals can give a high variability in replicate analyses. Grinding the laboratory sample finely should reduce the variability of the results.

Very little of the metal elements in the tea leaf gets into the tea infusion because most form insoluble complex compounds with organic compounds in the leaves.

We always seek to reduce the variability between duplicates to increase the precision of the determination. High variability between analytical duplicates would mean that real differences between samples from an experiment would be more difficult to detect.

4.4.2 Soil Sampling.   The types of land to be sampled should be distinct in some clear way – such as by soil colour, cultivation, slope, drainage or soil type. If it is available, a soil map of the area being sampled would give better definition of the samples taken.

In order to apply a statistical interpretation, we need replication of the identified types of land. Replicate areas do not have to be contiguous; they could be in separate fields but at a similar level, for example. The null hypothesis is that the soils from the different types of land are not different in available P content. A follow-up experiment could determine the P contents of vegetation associated with the soil sites.

Further replicate samples from within each sampling area could be used to determine the variability of soil within each type of land but that is not the objective of this investigation.

CHAPTER 3

Planning to Work in the Laboratory

1 MAKING ARRANGEMENTS

If you are planning to do some analysis yourself there are several steps that you need to take before you start work in the laboratory.

Having discussed with the responsible person in the laboratory what it is that you want or expect from the laboratory facilities, you must come to an agreement over what activities you will do yourself, how much supervision you will need and what work the laboratory is required to do for you. Charges, if any, must be agreed and approved in advance. The level of charges will inevitably have some effect upon the decisions made about the work that can be done and who is to do it.

Once you have settled these decisions, then:

- Check with the laboratory staff when it is possible or convenient for you to do the analytical work.
- Make sure you know the procedure that you will follow and ask the laboratory to give you a copy of it.
- Note the reagents and equipment required and ask the staff if they are all available – they may need to be ordered before the work can begin.
- It may be necessary to book an instrument to ensure that it is available for you when you need it.

Know your chemicals: it can be very easy to confuse one chemical with another, especially if their names and packaging are similar. Also, the same compound may have different forms – for example it may be an anhydrous powder or hydrated crystals. Mistakes at this stage are wasteful of time and reagents; and at worst may result in serious accidents. Always double check that you are using the right chemical.

The packaging of the chemical should tell you a lot about it, including the potential hazards. Additional information is provided in the manufacturers hazard data sheets (also called material safety data sheets – MSDSs). Other sources of information on chemical hazards are given later in this chapter.

From the point of view of personal safety it is useful to think of hazard and risk as two separate but connected subjects for concern:

- The hazard presented by a substance is its ability to cause injury or ill health.
- The risk presented by the substance takes account of the likelihood of the injury or ill health arising in practice.

We may express this as:

$$\text{risk} = \text{hazard} \times \text{probability}$$

So, if a hazardous substance is properly used, it should present only a small risk. Again, a hazardous substance, used in very small quantities, may pose only a very small risk. The aim must be to take steps to minimise the risks to yourself and others in using the chemical reagents.

In most countries there are regulations that an employer must, under law, follow to ensure the safety of employees and others that might be affected by the work being done. In the UK the use of chemicals is subject to a set of regulations called 'The Control of Substances Hazardous to Health' (COSHH) (Health & Safety Executive (1999)).

The COSHH regulations are concerned with the risk to the person of exposure to so-called 'controlled' substances. These include chemicals that are defined as very toxic, toxic, harmful, corrosive or irritant, as well as dust of any kind in substantial concentration in the air. In practice, the user also needs to know if a chemical in use has nil or minimal hazard, so the hazard information should be readily available for all chemicals used in the laboratory, not just the 'controlled' ones.

Fortunately the Health and Safety Executive (HSE) publishes each year a booklet called EH40, which lists the chemicals and dusts that are subject to control. The HSE defines two levels of control: the maximum exposure limit (MEL), which must never be exceeded; and the occupational exposure standard (OES) which is a realistic target for the workplace.

In accordance with the COSHH regulations the employer must:

- Ensure that substances are moved, stored and used safely.
- Give you the information, instruction, training and supervision necessary for your health and safety.

- Take proper precautions to prevent employees and others being exposed to substances which may damage their health. It also requires proper procedures to deal with spillages, escapes and disposal of substances in use.
- Provide, free of charge, any protective clothing or equipment required (safety glasses, gloves, laboratory coats, safety shoes).

Anyone acting in a supervisory or training capacity on behalf of the employer is personally responsible for ensuring that the regulations are observed.

However, as an employee (or trainee) you also have obligations under the law to:

- Take reasonable care for your own health and safety and that of other people who may be affected by what you do or do not do.
- Cooperate with your employer on health and safety.
- Not interfere with or misuse anything provided for your health, safety or welfare.

Therefore, if the work is allowed to proceed, the laboratory supervisor must provide all the facilities, information, protective clothing and equipment you need to work safely, but you must act responsibly on the information and make proper use of the safety equipment.

Usually the instructions for the analytical procedure that you follow will include information on the hazards associated with the reagents involved and recommendations regarding safety measures. Sometimes, especially if the procedure is one that the laboratory does not usually do, this information may not have been collated in advance.

In any case you should be aware of reliable sources of hazard data and safety recommendations:

- HSE booklet EH40, current issue.
- Warnings on the chemical bottle labels or packaging. All modern chemicals carry easily recognised warning signs stating TOXIC or CORROSIVE (see Figure 3.1), but with little other information. Old chemicals may not carry the warning signs but manufacturers will supply hazard data sheets on request.
- Catalogues from the chemical suppliers show for each chemical the warnings that are on the labels and also have an explanatory key for other symbols or numerical data provided (Figure 3.1).
- Material safety data sheets or hazard data sheets; these must be supplied with each chemical purchased. Unfortunately, many of

Information as it appears in chemical catalogues and on bottle labels:

Sodium dichromate GPR   MW 298.00

$Na_2Cr_2O_7.2H_2O$

Minimum assay (iodometric)   99%
Maximum limits of impurities
Chloride (Cl)   0.1%
Sulfate (SO_4)   0.5%

Sodium dodecyl sulfate

A mixture of sodium normal primary alkyl
sulfates, consisting chiefly of sodium dodecyl
sulfate
Assay (Epton titration)   90%

Symbols commonly used to show chemical hazards are:

 Corrosive (attacks and destroys living tissue including eyes and skin)

 Toxic (can cause death by swallowing, inhaling or by contact)

 Highly flammable (easily catches fire)

 Harmful (similar to but less dangerous than toxic substances)

 Oxidising (makes other materials burn more easily)

 Irritant (can irritate the eyes and skin)

 Dangerous to the environment (harmful to wildlife)

Figure 3.1 *Examples of chemical catalogue information and hazard symbols*

2 EXERCISES

2.1 Identification of Chemicals and Hazards and Assessing Risks

Information on making and using reagents for:

- The colorimetric determination of phosphorus in bone.
- The determination öf NDF (neutral detergent fibre) in animal feeds.

is summarised from a laboratory methods manual and reproduced below.

Read through the methods and note which chemical compounds are needed. If possible find the chemicals in the laboratory chemical store and note any hazard data printed on the container label. Also find hazard data from any of the sources listed above.

From the method summaries, see what you are required to do with each chemical and determine what risks you may face from using it:

- How much is used at one time?
- Is it volatile?
- Is it a fine powder?
- How may it invade the body? *e.g.* eyes, mouth, skin, breathing.
- What are its effects? *e.g.* toxic (poisonous, carcinogenic, mutagenic, or teratogenic), corrosive, irritant, short term or long term damage.
- Are you familiar with handling this chemical?

How can you minimise the risks to yourself and others in the laboratory? *e.g.* wear laboratory coat, safety spectacles, gloves, face mask, reduce the quantity handled, use a fume-cupboard.

How much do you need? The less you use, the smaller the risk is to yourself and others around you. Suppose you want to analyse 20 samples . . .

Can *you* work safely with these chemicals? Tabulate the hazards and safety measures as in Tables 3.1 and 3.2.

2.2 Colorimetric Determination of Phosphorus

A yellow coloured complex of phosphovanadomolybdate is formed in acid solution and the absorbance is measured at 420–440 nm.

2.2.1 Reagents. Note: The purpose of this exercise is to assess the risks associated with the preparation and use of the reagents, but not to make them.

- **Reagent 1**   0.05% Ammonium metavanadate in 3.5% nitric acid.
  Add 50 cm^3 of conc. nitric acid (70%) to about 800 cm^3 of water in a 1000 cm^3 volumetric flask. Weigh 0.5 g of ammonium metavanadate into a beaker and dissolve it in deionised water. Add it to the acid in the volumetric flask, make up to volume with water and mix well.
- **Reagent 2**   1.5% Ammonium molybdate.
  Dissolve 15 g of ammonium molybdate in about 300 cm^3 of deionised water. Transfer it to a 1000 cm^3 volumetric flask. Make it up to volume with deionised water and mix well.
- **Reagent 3**   Ammonium vanadomolybdate working solution.
  Mix together 300 cm^3 of the 0.05% ammonium metavanadate solution (Reagent 1) and 200 cm^3 of the 1.5% ammonium molybdate solution (Reagent 2).

2.2.2 Analytical Procedure (Summary).

- Take 5 cm^3 of your sample solution and add 5 cm^3 of the ammonium vanadomolybdate working solution (Reagent 3).
- Mix well and after 15 minutes read the absorbance at 420 nm.

Table 3.2 *Risk assessment for colorimetric determination of P*

Chemical	Hazards	Safety measures
ammonium metavanadate	info from bottle: from other sources:	
ammonium molybdate	info from bottle: from other sources:	
nitric acid 70% (*i.e.* conc.)	info from bottle: from other sources:	

2.3 Determination of Fibre

Neutral detergent fibre is a descriptive term for some of the insoluble matrix of plant cell walls. It consists of cellulose, hemicellulose, lignin and some fibre-bound nitrogen and minerals. Other fractions of fibre are described as crude fibre, acid detergent fibre and acid detergent lignin. NDF is the dried residue remaining after hot extraction of a sample with a detergent solution of sodium dodecyl sulfate.

2.3.1 Reagents.   Note: The purpose of this exercise is to assess the risks associated with the preparation and use of the reagents, but not to make them.

- **Reagent 1**   Neutral detergent solution.
  Dissolve separately in hot, deionised water:
  18.6 g diaminoethanetetraacetic acid, disodium salt
  6.81 g disodium tetraborate
  4.56 g disodium hydrogen orthophosphate, anhydrous
  10 cm^3 triethylene glycol (Trigol)
  In the fume-cupboard, weigh 30.0 g per litre sodium dodecyl sulfate (sodium lauryl sulfate) into a 1 litre beaker. Dissolve it in about 500 cm^3 water and add the other dissolved reagents.
    Cool the solution and, using a pH meter, adjust the pH to 6.9–7.1 with 1 M sulfuric acid or 1 M sodium hydroxide. Transfer the solution to a volumetric flask and make it up to 1 litre with water.
- **Reagent 2**   Amylase working solution.
    From heat stable amylase (Sigma 3306) solution, take 2 cm^3 and add it to 80 cm^3 of cold NDF solution.
- **Reagent 3**   Acetone – laboratory reagent grade.

2.3.2 Using the Reagents (Summary).

- Weigh 1 g sample into a 500 cm^3 conical flask, add 100 cm^3 NDF solution (Reagent 1) and 4 cm^3 of amylase working solution (Reagent 2).
- Boil under reflux for 1 hour.
- Filter while hot into a weighed, sintered glass crucible and wash thoroughly with hot water to remove all the detergent.
- Finally wash once with acetone.
- Dry the sample at 100 °C for 4 hours, cool in a desiccator and weigh again.

3 SAFETY IN THE LABORATORY

3.1 Code of Practice

Every laboratory should have a code of practice. This will usually identify the laboratory supervisor as the person responsible for the safety of persons within the laboratory. Therefore the laboratory supervisor will decide on such matters as to whether the ability, conduct and behaviour of persons within the laboratory is conducive to the maintenance of a safe environment there. Self discipline is expected, but if it is not evident then discipline has to be imposed.

  The code of practice will cover such matters as:

- Trainees and students must work under the supervision of laboratory staff.

Table 3.2 *Risk assessment for the determination of NDF*

Chemical	Hazards	Safety measures
sodium dodecyl sulfate	info from bottle: from other sources:	
diaminoethanetetraacetic acid, disodium salt	info from bottle: from other sources:	
disodium tetraborate	info from bottle: from other sources:	
disodium hydrogen orthophosphate, anhydrous	info from bottle: from other sources:	
triethylene glycol	info from bottle: from other sources:	
amylase	info from bottle: from other sources:	
acetone	info from bottle: from other sources:	
sodium hydroxide 1 M solution	info from bottle: from other sources:	
sulfuric acid 1 M solution	info from bottle: from other sources:	

- All persons working in the laboratory must be made aware of the COSHH assessment for the procedures that they are to follow and the recommended safety measures must be adopted.
- All persons must at all times take care to protect themselves and colleagues when working in the laboratory.
- No persons may do analytical work when they are alone in the laboratory (at lunchtime or outside normal working hours).
- All coats and bags must be stored tidily in a designated area and must not obstruct access.
- Laboratory coats and other appropriate personal protective equipment, such as safety spectacles and face masks, must be worn as necessary and removed when leaving the laboratory.
- No food, drink or tobacco is permitted in the laboratory.
- All operations involving concentrated acids, solvents and toxic substances must be done in the fume-cupboards.
- All reagent solutions must be clearly labelled with the name and concentration of the reagent and the name of the person responsible for it.
- Pipetting must be done by means of pipette fillers or automatic pipettes; never by mouth.
- All spillages must be cleared up, and wet floors dried, immediately.

- For disposal, only low-toxicity water soluble waste may be flushed down the sinks (with copious amounts of water), other chemical waste must be kept in waste bottles.
- Chlorinated solvents must never be mixed with, or added to, other solvents.
- Wash hands before leaving the laboratory.

3.2 Dangers in the Laboratory

There is a wide variety of potential dangers in the laboratory (Table 3.3) so try to keep 'alive' to them all.

Table 3.3 *Some sources of danger in the laboratory*

Carelessness	Falling objects	Machinery
Chemicals	Fire	Noise
Cleaning agents	Fumes	Open drawers
Compressed air	Glassware	Open cupboards
Confined spaces	High level shelving	Sharp edges
Electricity	Hot or cold surfaces	Slippery floor
Electricity and water	Ladders	Solvents
Explosions	Lifting and handling	Vacuum

3.3 Examples of Bad Bench Work

Following the first example given in Table 3.4, comment on the situations listed in Table 3.5, which put the staff or the analytical work, or both, at risk.

Table 3.4 *An example of the danger of poor laboratory working practice*

Unsafe situation	Risk	Reducing the risk
Example: pipette standing in a volumetric flask	flask could easily be knocked over and spill its contents	always stopper containers to prevent spills (and contamination); the pipette should be placed safely on the bench or tray

Table 3.5 *Other examples of poor laboratory working practices*

1	Chipped beaker in use
2	Flask with ill-fitting stopper
3	Glass pipette with broken tip
4	Plastic pipette tips scattered on the bench
5	Unlabelled samples on the bench
6	Sample spilt on the bench
7	Broken electrical plug
8	Electric lead able to fall into a sink
9	Food, dirty mug and cigarette ends on the bench
10	Cupboard and drawers left open
11	Flask stoppers left on the bench
12	Greased desiccator lid left flat on the bench
13	Reagents labelled with no date or concentration
14	Plastic glove left inside out
15	Automatic pipettor propped upside down
16	Reagent spilt on bench
17	Top off wash bottle and left on bench
18	Weighings written on scrap paper
19	Torn lab coat thrown on bench
20	Balance pan and surrounding area dirty
21	Filter paper box left open
22	Tops left off reagents
23	Broken glass on the bench
24	Broken glassware in the waste paper basket
25	Used first aid plaster left on the bench
26	Written procedure with illegible alterations
27	Crumpled and dirty paper towel left on the bench
28	Hot plate left on but not in use

CHAPTER 4

Weights and Measures

1 INTRODUCTION

Weight and volume measurements form the basis for almost all analytical work we do. Eventually, for some determinations, we will have to make some other measurements too, but to turn these measurements into useful results, our calculations most frequently include weight and volume parameters. Often we want to express our result as a concentration – for example as *milligrams* of sodium per *kilogram* of cheese *(ppm)*; or *grams* of calcium per *100 grams* of cereal *(% w/w)*. Or the results might be more useful on a volume basis such as *micrograms* of benzene per *litre* of bottled water *(ppb)* and *milligrams* of alcohol per *100 millilitres* of blood *(parts per 100,000 w/v)*. The use of the older common names for concentration units (percent, parts per million *etc.*) is being discouraged because their meaning can be interpreted in different ways. Scientific units are preferred and more acceptable for laboratory work (see Table 4.1).

Table 4.1 *Units expressing concentration*

Scientific units		Common name
		(now avoided as scientific terms)
Liquids	*Solids*	
g 100 cm^{-3}	g 100 g^{-1}	percent (%)
g l^{-1}	g kg^{-1}	parts per thousand ($^0/_{00}$)
mg l^{-1}	mg kg^{-1}	parts per million (ppm)
μg l^{-1}	μg kg^{-1}	parts per billion (ppb)
ng l^{-1}	ng kg^{-1}	parts per trillion (ppt)

With our usual methods of analysis we cannot directly measure the things we want to in solid materials such as wood, stone, crops, soil, toys, plastic *etc.* We often have to get the components we are interested in into solution before we can determine the amounts present. At each stage we must know and record:

41

- What amount (weight) of the material we take as a test sample.
- What volume of solution we make from the test sample.
- How much of the solution (volume) we take for the determination.
- What is the result of the determination in solution (often a concentration such as $mg\,l^{-1}$).

Then we can calculate back, to work out the amount or concentration of the measured component within the material we started with. Our measurement of weight or volume at each stage must be good enough to allow us to express the final result with the accuracy and precision appropriate to the original purpose intended.

2 ACCURACY AND PRECISION

In an archery competition two contestants share the same equipment, but contestant A is more experienced and skilful than the other, B. After the first five shots each, they readjusted the sights to align them correctly (Figure 4.1).

Mark which of the results were:
A (1st shots);
B (1st shots);
A (2nd shots);
B (2nd shots).

Mark which shots showed:
Good accuracy and poor precision?
Good accuracy and good precision?
Poor accuracy and poor precision?
Poor accuracy and good precision?

Similarly, we have to be concerned with the accuracy and the precision of the measurements we make in analysis. If the instrument or the balance has been calibrated using incorrect standard weights, our results will be inaccurate, however carefully we do the work. Again, if the volumetric flask is not correctly calibrated, the measurements we make with it will be wrong. Incorrect calibration gives systematic or non-random errors in the results. If the error is known and is consistent, we may be able to correct our results by multiplying by a factor. However, it is better to eliminate the systematic error at its source if we can, then we know that the instrument or method in use is correctly set-up to give results that are accurate within a specified range of the 'true' result.

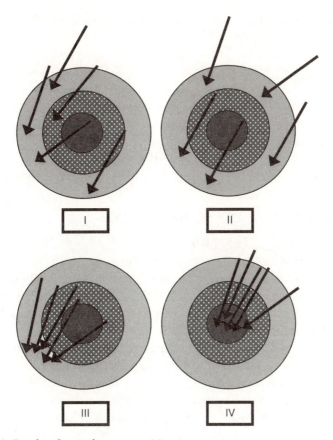

Figure 4.1 *Results of an archery competition*

Almost all measurements are subject to some random error. This is the inherent variability of the process of measurement. Even something as apparently straightforward as measuring the length of a table with a tape measure will have some variability. At some level, repeated measurements will not be identical. If the readings are unbiased they will be evenly spread around the mean result and, if the tape measure is *accurate*, the mean result of a set of replicates will be very close to the *'true'* result.

In analytical work a major aim is to *increase the precision* of our measurements. This means *reducing the variability* of replicate measurements as much as we can, or as much as is necessary for the purpose of the measurement. The measuring instruments should be set to be responsive to the parameter we are measuring, but relatively unaffected by other extraneous influences. In analysis, adequate precision of the measurement will depend not only on the instruments used but also on the technical skills of the analyst and standardisation of the analytical procedure.

2.1 Note on this Section

We should expect that archer A, as the more experienced and skilful, would shoot with greater precision *i.e.* her shots would be grouped closer together (III and IV) and her accuracy improved after re-adjusting the sights (IV). Similarly B's accuracy improved after re-adjusting the sights but his shooting remained less precise (I and II).

3 WEIGHING AND CARE OF BALANCES

Although we speak loosely of the weight of an object, weight is really the force of attraction exerted by earth's gravity on the object's mass. Gravity is not constant all over the earth and it decreases with altitude *i.e.* with the distance from the centre of the earth. So the weight of an object varies slightly from place to place on the earth, although its mass does not. On the moon, astronauts have the same mass as they do on earth but their weight on the moon is reduced to one sixth. This is because the mass of the earth is 82 times, and its diameter is about 3.8 times, that of the moon. Gravity at the surface of a spherical body is proportional to its mass and inversely proportional to the square of its diameter. So the gravity ratio of earth:moon is $82:(3.8)^2$ or 5.7:1.

A semi-micro balance that measures accurately a mass of 200.0000 g at ground level will show only 199.9974 at one floor up (4 m).

In older mechanical balances we simultaneously compared the weight of the object we were weighing with the weight of standard masses under the same gravitational conditions. So adjustment to local conditions was made automatically in the weighing procedure.

In electronic balances, we do not counterbalance the object simultaneously with a standard mass. Instead, the scale pan is linked to an electronic device that gives a direct digital display of the weight reading. The balance must be accurately calibrated before we weigh the object, and the calibration has to be 'adjusted' to the local gravitational field.

Modern electronic analytical and precision balances are extremely reliable. Once the balance has been correctly calibrated, the only adjustment that the user normally has to make is to the levelling mechanism. This is done by raising or lowering the adjustable foot screws to centralise the bubble in the spirit level device. The bubble should be checked each time the balance is used and the level must be readjusted whenever the balance is moved, even slightly. This adjustment is important because a mass of 200 g will weigh 199.9975 g if the balance is tilted by only 0.29°; a loss of 2.5 mg.

The user can set the display to zero (tare) with any weight on the pan, up to the maximum; adjust the degree of damping (stability); and select the weighing units required (although in the laboratory we invariably use grams). The calibration of balances is done with standard masses made of stainless steel with a density of 8.0 g cm^{-3}.

Air buoyancy – this affects the weight of everything that has a different density. Only in very specialist work is any allowance made for the air buoyancy effect on weighing, so for practical purposes this inherent error is universally accepted in analytical chemistry laboratories.

Location – modern electronic balances have been designed to give reliable results under difficult conditions. However, they may be affected by vibration, draughts and temperature changes. So far as possible, a balance should be located to avoid or reduce these effects.

Electrostatic effects – glass and plastic containers and powdered samples can become charged electrostatically. This causes attraction or repulsion between the object and the balance. The effect slowly dissipates and so the weight reading changes. The static charge may be reduced by placing the object in a metal container or allowing it to equilibrate in a humid atmosphere. Some modern balances have built-in antistatic devices.

Stability of samples – the precision and accuracy of a weight measurement will be impaired if the sample changes during the weighing process. A common effect is for dry samples to pick up moisture or wet samples to lose moisture while being weighed. Samples which have been dried in an oven should be cooled in a desiccator containing dry silica gel (self-indicating silica gel is blue when dry and pink when it needs drying) and kept there until immediately before being placed on the scale pan. Moist samples should be kept in closed containers.

Routine care of the balance:

- Always treat the balance carefully.
- Weigh powders or chemical reagents onto weighing paper or into a container, never put them directly onto the pan.
- Do not drop heavy objects onto the pan.
- Do not exceed the maximum weight accepted by the balance.
- Always remove powder or moisture from the pan immediately.
- Clean the pan and sweep out the weighing compartment thoroughly before and after use.
- Clean the bench area around the balance before and after use.

- Keep the balance covered when it is not in use to protect it from dust.
- Electronic balances benefit from being left on continuously. This keeps the balance warmed up, reducing the adverse effects of atmospheric humidity within the balance. It also avoids the power fluctuations associated with switching on and off which can shorten the life of some components. If the balance is switched off overnight, switch it on at the start of the working day and allow it to go through its initialisation programme.

In weighing:

- Cleanliness in, on and around the balance is essential.
- Use a small vessel to reduce air buoyancy effects.
- The balance and the sample should be at the ambient temperature because the movement of air by convection makes cold objects appear heavy and hot objects appear light.
- Be aware of problems due to electrostatic charges.
- Avoid touching the sample, especially for high resolution work – a single finger print can attract moisture and add 400 µg.
- Hygroscopic samples should be weighed in a closed container.

4 CARE AND USE OF GLASSWARE

Glass is widely used in the laboratory because:

- It is resistant to chemical attack.
- It is transparent.

We use a large range of glass apparatus in teaching, research and industrial laboratories. Only a few common examples are listed in Table 4.2.

Table 4.2 *Examples of common laboratory glassware*

test tubes	beakers	watch glasses	pipettes
sample tubes	measuring cylinders	dispensers	burettes
sample jars	syringes	volumetric flasks	condensers
reagent bottles	conical flasks	desiccators	retort flasks

The glass commonly used for making bottles is soda glass, but much of the glassware used in the laboratory is made of borosilicate glass, which has several advantages, mostly due to its much lower coefficient of thermal expansion. Borosilicate glass:

- Can be safely used up to 600 °C.
- Is less liable to failure when suddenly heated or cooled.
- Glass apparatus can have a greater wall thickness and so has greater mechanical strength without affecting the thermal strength.
- Also has greater resistance to chemical attack and is less liable to surface deterioration with age.

4.1 Cleaning Glassware

Cleaning is much easier when the residue is new, so glassware should be cleaned soon after use. Often washing with warm soapy water, using a detergent such as 'Teepol' or 'Decon 90', is sufficient. If a brush is used, make sure that metal parts of the brush are covered to avoid scratching the glass.

Organic material, persistent and greasy residues may be removed by solvents or by the use of dichromate/acid cleaning mixture, but note the chemical hazard warnings in making and using the mixture.* Apart from the danger from the concentrated acid, chromium is highly toxic and volatile chromium compounds are carcinogenic.

As the final stage of washing, rinse the glassware thoroughly with distilled or deionised water. Leave to drain-dry or dry off in an oven. Remember that volumetric glassware should not be heated above 60 °C, because the process of expansion on heating is not strictly reversible when the glass contracts on cooling, so the volume may be permanently affected and the original calibration be made inaccurate.

4.2 Ground Glass Joints and Stopcocks

Keep all ground glass joints, including glass stopcocks, completely free from dirt and lightly lubricated with Vaseline or stopcock grease. Grit in a joint may cause leakage or even damage. Therefore the joints should be wiped clean and relubricated before use. Glassware used for trace analysis needs special precautions to avoid contamination.

Desiccators (Figure 4.2) often suffer from misuse. The main points in using a desiccator properly are:

---

* The dichromate/acid cleaning mixture is made by dissolving about 60 g of sodium dichromate per 1 litre of concentrated sulfuric acid. Note the hazard data for these chemicals and use full personal protection, especially to hands and eyes when making or using the cleaning mixture. It is effective in removing most residues, grease and algae from flasks, burettes and pipettes if left in contact with the glass overnight. The mixture should be filtered occasionally through glass wool. Another good degreasing agent is 100 g KOH in 50 cm^3 water then made to 1 litre with industrial alcohol.

- Hold the lid only by the collar; never by the stopper which may suddenly loosen, so that the lid can drop catastrophically, resulting in possible injury and a cost to the laboratory of £150 – 200.
- Clean and lightly grease the ground glass flanges of the base and lid, so that they can maintain an air-tight seal.
- The lid must not be put flange-down on the work surface or else dirt and grit will stick to the greased flange.

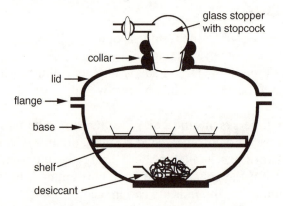

Figure 4.2 *Desiccator*

- Keep the desiccant (usually silica gel) in the oven at 100 °C when it is not in use.
- Put hot oven-dried or ashed samples into the desiccator at about 100 °C. Open the stopcock and fit the lid carefully onto the base. After about one minute, close the stopcock. This prevents the expanding air in the desiccator from pushing the lid or the stopper off. With the stopcock closed, the cooling air inside contracts. The reduced air pressure keeps the desiccator sealed from the outside air and humidity.
- To remove the lid, open the stopcock slowly, carefully slide the lid firmly to one side and lift it off by the collar.

4.3 Safety

There is always a safety concern in handling glassware. Beakers, flasks and any other items with chipped rims should not be used – mainly because the chipped region is sharp and presents a danger during use and washing up. If the item is worth saving then the chipped area should be filed smooth or annealed in a flame.

Whenever you are manipulating glass make sure your hands are properly protected. Be careful when fitting a pipette filler to, and removing it from, a pipette. If you need to remove old tubing or rubber bungs from glassware they should be cut off rather than pulled or twisted. It is better to sacrifice the rubber items rather than your hands.

4.4 Volumetric Glassware

Volumetric glassware is marked with one or more graduation marks to indicate liquid volumes. Calibration of glassware is usually at 20 °C but in the USA sometimes calibration is at 25 °C.

4.4.1 Effect of Temperature. Expansion does not have much effect on the borosilicate glassware, but it does on the water contained. So, in calibrating glassware with water by weight, correction should be made for temperature.[†]

For normal laboratory work the error due to temperature is not very large at ambient temperatures (approximately 0.25% per 10 °C difference from the calibration temperature). So solutions should be at ambient temperature before being measured volumetrically.

4.4.2 Rough Accuracy. Beakers, conical flasks, reagent bottles, sample tubes and test tubes may have graduation marks, but these are only for rough guidance and their accuracy is unreliable (Figure 4.3).

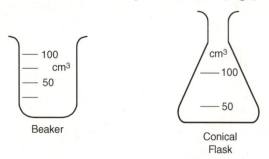

Figure 4.3 *Laboratory glassware with graduation marks for guidance only*

4.4.3 Moderate Accuracy. Measuring cylinders are calibrated with only moderate accuracy. They are usually expected to give within 2% of the nominal marked volume.

4.4.4 Good Accuracy. Graduated pipettes and burettes should deliver within 1% of the nominal marked volume. There are several different types of calibration of graduated pipettes:

† Volume (cm³) of 1 g of water at various temperatures (°C):

temp.	vol.	temp.	vol.	temp.	vol.	temp.	vol.
10	1.0013	16	1.0021	22	1.0033	28	1.0047

type 1   delivers measured volume from zero at the top to selected graduation mark;

type 2   delivers measured volume from selected mark to zero (the tip) – so the final drop must be removed or blown from the pipette, unlike the usual procedure for using a transfer pipette (see Section 4.4.6);

type 3   calibrated to contain a given capacity from the tip to a selected graduation mark *i.e.* to remove the selected volume of solution.

4.4.5 Highest Accuracy.   The volumetric glassware items with the highest accuracy are bulb pipettes and volumetric flasks (see Figure 4.4).

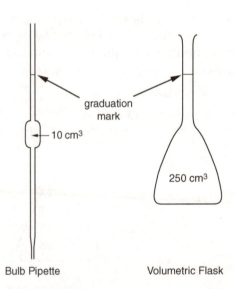

graduation mark

10 cm^3

250 cm^3

Bulb Pipette   Volumetric Flask

Figure 4.4 *Laboratory glassware with graduation marks of the highest accuracy*

The ordinary glass bulb pipette is a type of 'transfer' pipette. To meet specifications of Class B, pipettes must conform to **BS 1583** and are colour coded.

capacity		5	10	25	50	100 cm^3
tolerance	±	0.01	0.04	0.06	0.08	0.12 cm^3

For Class A pipettes, tolerances are about half those shown for Class B.

4.4.6 Correct Use of the Glass Pipette.   Transfer of the correct volume depends upon the proper use of the pipette:

- Make sure that the pipette filler is dry and that any solution previously used does not contaminate your sample. Fit it to the pipette and check that it is working properly *i.e.* it draws solution into the pipette and the pipette does not leak when held vertically. Do not force the filler onto the pipette in trying to stop a leak – the pipette is likely to break and cause an injury.
- Rinse the pipette with the solution it is to contain.
- Fill the pipette again, so that the solution is above the graduation mark and has no bubbles in it.
- Wipe the outside of the pipette with a tissue or paper towel.
- Adjust the solution correctly to the mark (allowing the bottom of the meniscus to sit upon the graduation mark, with the pipette held vertically and viewed with the eye in line with the mark, Figure 4.5).

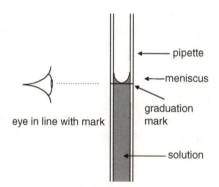

Figure 4.5 *Viewing the liquid level in a pipette*

- Touch the pipette tip against the inside of the vessel from which the solution was taken to remove any drop of solution remaining on the outside of the tip.
- Release air into the top of the pipette (with some pipette fillers it is necessary to remove the filler from the pipette at this stage) and let the solution drain naturally (by gravity only) into the collecting vessel.
- Hold the pipette vertically for five seconds after the last drop.
- Touch the tip against the inside of the vessel – this removes some of the solution held in the tip. The final portion of solution remaining in the tip should not be expelled, because the calibration of the pipette will have allowed for it.
- Wash the pipette so that solution does not dry in it.

Do not:

- Blow down the pipette.
- Hold it by the bulb (hand warmth will alter the volume).
- Allow solution to dry out in the pipette tip.

5 PRACTICAL EXERCISE

5.1 Measurement Uncertainty

In your notebook, prepare a worksheet like the one shown in Table 4.3. For this exercise you need:

- Analytical balance reading to 0.0001 g (0.1 mg).
- 25 cm^3 glass bulb pipette (Class B).
- 100 cm^3 beaker.
- Thermometer 0 to 100 °C, reading to 0.1 °C.
- Distilled or deionised water (about 200 cm^3) in a conical flask.
- Paper towel.

Note that you cannot check the accuracy of the pipette unless you know that the balance and the thermometer are accurate too. However, you can estimate the precision with which you can measure 25 cm^3 with the pipette. If the balance is correctly calibrated and set-up properly, the random error due to weighing will be very small compared with that due to pipetting.

Dissolved air in the water affects its density. De-gas the water – the usual techniques are by boiling, using an ultrasonic tank or bubbling helium through it – about 10 minutes for 1 litre.

Inspect the pipette and the pipette filler. Ensure that the pipette is not chipped and that the filler is clean and dry, otherwise the volume measurements will not be correct. Fill the pipette with water and allow it to drain out. Check that the tip is not blocked and the water drains freely. If the pipette is clean no droplets of water will be left on the sides of the pipette. If the pipette is dirty or blocked, clean it as advised by the laboratory supervisor or select another one that is clean.

a) Ask the laboratory supervisor to make the checks normally done in the laboratory to ensure that the balance is working satisfactorily.
b) Check and record the temperature of the water.
c) Make sure that the beaker is clean and dry, put it on the scale pan and tare the balance to zero.

Table 4.3 *Worksheet for exercise on measurement uncertainty in pipetting*

Date		
Balance checked and found OK	yes	no
Pipette checked for cleanliness and absence of chips	yes	no
Pipette filler function checked and found OK	yes	no
Thermometer function checked and found OK	yes	no

Results:

	weight of nominal 25 cm^3 water by pipette	calculated volume (using mean cm^3 g^{-1})
(i)	_____ g	_____ cm^3
(ii)	_____ g	_____ cm^3
(iii)	_____ g	_____ cm^3
(iv)	_____ g	_____ cm^3
(v)	_____ g	_____ cm^3
(vi)	_____ g	_____ cm^3
(vii)	_____ g	_____ cm^3
(viii)	_____ g	_____ cm^3
(ix)	_____ g	_____ cm^3
(x)	_____ g	_____ cm^3

Water temperature:
Start _____ °C
End _____ °C
Mean temp _____ °C
Mean cm^3 g^{-1} _____
(see footnote [†] on page 49)

Total _____ cm^3

Standard deviation (s) ± _____ cm^3 (for an individual measurement)

Mean _____ cm^3

Uncertainty is expressed as standard deviation or relative standard deviation (%) (= standard deviation × 100/mean)

The manufacturer's uncertainty for the 25 cm^3 pipette (m) = ± ____ cm^3

Combined uncertainty = $s + m$ = ± ____ cm^3

d) Use the pipette in the recommended manner (see Section 4.4.6) to transfer 25 cm^3 of water from the conical flask to the beaker.

e) Return the beaker to the balance and note the weight.

f) Empty the water from the beaker back into the conical flask, dry the beaker with a paper towel.

g) Repeat the steps c–f to obtain 10 weighings.

h) Check and record the temperature of the water again.

i) Calculate the volumes of the 10 pipettings, their mean and standard deviation.

The standard deviation is the 'standard uncertainty' of the measurement for this pipette and the relative standard deviation gives the standard uncertainty for the pipette expressed as a percentage of the mean.

To calculate a general value for pipettes of this size we need to add the uncertainty associated with the manufacturing tolerance limits (see Section 4.4.5). The manufacturer's uncertainty is: tolerance/$\sqrt{3}$.

For a 25 cm^3 Class B pipette this is 0.035 cm^3 (*i.e.* 0.06 cm^3/1.73).

6 VALIDATING THE ANALYSIS

6.1 Are We Sure that the Results are Correct?

If we know that the balance has been correctly calibrated, checked before use and used properly, then there will be no systematic bias away from the 'true' result in the weighings made on it. The average of several weighings of a single stable object, such as a standard mass, will be very close to the 'true' weight, so the balance weighs accurately.

At some level of measurement there may be some variation in repeated weighings but we expect this to be extremely small compared with the weight itself – so the balance weighs with good precision as well as good accuracy.

6.1.1 Combined Uncertainty.   Usually it is not possible to do a number of replicates of each test result in order to determine its uncertainty. However the uncertainty associated with different stages of the test procedure can be determined in advance and an overall combined uncertainty estimated for the test method.

The practical exercise with the pipette shows how the uncertainty due to one component of a test method could be measured. In this case we expect measurements made with both the balance and the thermometer to be both accurate and precise. Most of the uncertainty in the measurement of the volume of repeated pipettings of water would arise from the pipetting technique.

In a test procedure there may be uncertainty due to several volume measurements by pipette or volumetric flask that could affect the final result. In addition there may be an uncertainty due to reading the result on an instrument, such as a colorimeter or an atomic absorption spectrometer.

When the major components contributing to the overall uncertainty are known it may be possible to reduce some or all of them, for example by:

- Changing the sample plan to reduce variability between field replicates – for example by stratified sampling or by increasing the number of increment samples that make the composite sample.
- More thorough mixing, homogenising or finely grinding the laboratory sample.
- Improving laboratory skills and performance in pipetting and liquid handing.
- Selecting and keeping particular items of equipment for the job – pipettes, pipette fillers, volumetric flasks.

• Optimising instrument settings for better precision.

The standard deviation of a measurement is its standard uncertainty. We can also estimate a component uncertainty. For example in making an extract up to volume in a volumetric flask, perhaps we can judge the volume at the graduation mark ±0.10 cm³ (two drops). So the estimated uncertainty in that case is ±0.10 cm³.

If the uncertainty from one step of the procedure is dominant, that is taken as the overall uncertainty. If there is not a dominant component, the main uncertainties, expressed as relative standard deviations (to eliminate units) can be combined to arrive at an overall uncertainty. This is done by taking the square root of the sum of their squares:

$$s_t = \sqrt{(s_1{}^2 + s_2{}^2 + \ldots + s_n{}^2)}$$

where s_t is the overall uncertainty. We did not use this method of combining uncertainties in the exercise on measurement uncertainty in Section 5.1, but simply added the manufacturer's uncertainty to the uncertainty we found in using one pipette to obtain a general uncertainty for any Class B pipette of that volume.

We expect the random variation of replicate results to have a normal distribution. In a normal distribution:

67% of the values fall within ± s of the mean;
95% of the values fall within ± $2s$ of the mean;
99% of the values fall within ± $3s$ of the mean.

So the combined standard uncertainty defines a range of values around our result, but we have only 67% confidence that the 'true' value is in this range. Multiplying the combined standard uncertainty by a factor of two defines a range within which we can say there is 95% confidence that the 'true' result lies. The factor of two is the 'coverage factor' and the new range is the 'expanded uncertainty'. A coverage factor of three would give a range within 99% confidence limits.

To make the user aware of the situation, a laboratory may report the result in the form:

measured value = 57.2 (units)
uncertainty of measurement = ± 0.2 (units)

The reported uncertainty is an expanded uncertainty based on a standard uncertainty of 0.1 units multiplied by a coverage factor of two, which provides a confidence level of approximately 95%.

6.1.2 Check Samples and Proficiency Testing. It is clearly useful to have an indication of the reliability of the result and the potential range within which it could fall, but the uncertainty is not always reported. A laboratory nevertheless must satisfy itself that the procedures in use give results that are generally acceptable. This it may do by including check samples in each batch of analysis together with the new samples for testing and the calibration standards for the instrument. So each batch of analysis may consist of:

- A set of calibration standards – to assess or set-up the response of the instrument to the parameter being measured.
- Blank samples – consisting of all the reagents taken through the analytical procedure but without the inclusion of sample material – to assess the instrument response to the reagents in use with the current samples.
- New samples for testing.
- Some replicates of the new samples – to check the variability of the determination within the batch.
- Previously-analysed samples such as a laboratory bulk sample (or in-house check samples) – to check the variability of the determination between different batches.
- Certified samples (*i.e.* certified reference materials) may be used instead of, or together with, the other check samples – to check that the analytical procedure is capable of producing results of acceptable accuracy (information on a range of reference materials may be found on the LGC web site, *http://www.lgc.co.uk/ref.asp*).
- Further calibration standards spread throughout, and at the end of, the batch – to monitor any changes in the sensitivity of the instrument during the analysis.

The laboratory may also join a proficiency testing scheme in which a number of samples are analysed by many different laboratories and the results tabulated together for comparison. This shows whether or not the laboratory's results are in line with those from other laboratories doing similar work and so highlights where problems exist. It is particularly useful if certified samples are not available for a certain determination, or are too expensive for frequent use. See *http:// ptg.csl.gov.uk/schemes.cfm* and *http://www.lgc.co.uk/pts_schemes.asp*.

6.1.3 Laboratory Accreditation. The quality control routine noted above seems quite impressive. However, a laboratory's judgement of acceptability of its own work may be highly subjective. The quality of analysis may decline unless some measurable limits are imposed and monitored using independent nationally or internationally recognised criteria. One of the most suitable standards for laboratory accreditation is the ISO/IEC 17025 standard.

The main point about accreditation is that the laboratory must be able to satisfy an independent outside official body that the work it does is consistently of a quality appropriate for the purpose intended. So, a laboratory that is accredited for some or all of its work, is subject to regular inspections by the accreditation agency. The laboratory has to keep extensive documentation to show that all the requirements of the specifications for accreditation are met on a continuous basis. For example, it has to keep up to date evidence of staff competence and training, performance of equipment and the test results for reference samples (see the UKAS website *http://www.ukas.com*).

6.1.4 Traceability. One of the requirements of laboratory accreditation by the United Kingdom Accreditation Service (UKAS) is that all measurements necessary for the proper performance of a test should be traceable, where possible, to national standards of measurement. In practice this means that some laboratory equipment such as a balance, thermometer or pressure gauge, for example, must be regularly calibrated by an accredited specialist calibration company, if accurate weight, temperature or pressure measurements are necessary for the test.

The laboratory has to be able to check that the calibrated equipment is still within the specifications required each time it is used. It also has to check other similar items of equipment in the laboratory against the officially calibrated ones. So, for balances, it would keep some special check weights and record their weights each time the balance is used. Volumetric glassware must be of a recognised specification (such as BS 1583) and be checked regularly, in the manner of the practical exercise in Section 5.

6.2 Exercise

Do this exercise before turning to the notes that follow:

Suppose that you represent a company requiring essential analytical work upon which important financial decisions depend. You might have confidence in using a testing laboratory accredited by an officially recognised agency such as UKAS.

List some points that you would expect accreditation to require of the laboratory in regard to the work it does for your company. How would you convince yourself that you can trust the results from this laboratory?

6.2.1 Notes on the Exercise.   The company should expect that accreditation would ensure that:

- There would be a regular and independent assessment of the technical performance of the laboratory – by assessors appointed by the accreditation agency and the first assessment would take place before accreditation is granted.
- The staff making analytical measurements would be both qualified and competent to undertake their tasks – and the laboratory would keep records to show that staff had received relevant training and were competent.
- The laboratory would have well defined quality control and quality assurance procedures, with regular internal checks to see that the procedures were being followed.
- Analytical measurements would be made using methods and equipment which had been tested to ensure they were fit for the purpose. These would be officially recognised methods or laboratory methods which were validated by analysis of certified samples or interlaboratory comparisons.
- Analytical measurements made in one location would be consistent with those made elsewhere – the laboratory would have evidence that its results were consistent with those from another similarly accredited laboratory.
- Analytical measurements would be made to an agreed standard – the laboratory would be able to give confidence limits to the results and the company (*i.e.* the user) could judge if this would be acceptable.

CHAPTER 5

Digestion and Extraction

1 MAKING SOLUTIONS FROM THE SAMPLES

With the analytical instruments that most laboratories have, it is necessary to get the elements and other components that we want to determine into solution. The ways of making solutions from the samples are:

- Acid digestions – effectively dissolving the whole sample, or most of it.
- Dry-ashing at 450–550 °C to destroy organic matter, followed by dissolution of the residue in acid.
- Extraction – by hot or cold reagent solutions or solvents.

Acid digestion and ashing are usually used to determine the total contents of a component, such as a metal element, being measured. On the other hand, extractions may be targeted at the removal of a fraction of the component. Extraction may be repeated several times to remove all of the particular fraction from the sample; that would be an exhaustive extraction. In other procedures, a single extraction, under specified conditions, gives an indication of the amount of the component in the sample. The choice of the actual method used will depend on the purpose of the study. For example, it may be that the total content of a metal element in a plant indicates the requirement for that element to be added as a nutrient. Or, the total content may affect some aspect of further processing or a quality factor such as colour, taste or flavour. On the other hand an extraction may be more appropriate to determine the loss of the element in processing or cooking. In an example of practical work given in this chapter, a hot water extraction is used to measure the quantities of some metals that are released from tea leaves to be consumed in a cup of tea.

Before proceeding with the practical work, it is a good idea to refer back to the sample plan (Chapter 2) so that it is clear which samples need to be analysed to provide the information required. It is also recommended to read through the procedure carefully before starting the analytical work.

Take care: where the procedure involves handling strong acids and hot apparatus and solutions, make sure you have the appropriate training, equipment and protection to do the work safely (see the COSHH assessment in Chapter 4 and the reagent safety data in the Appendix). If in doubt, ask for help.

In this chapter we will explore the analysis of:

- Total contents – Mn and Zn in tea leaves by ashing and acid digeston (Section 1.1).
- Extractable contents:
  - hot water extractable Mn and Zn in tea leaves (Section 1.2), and
  - 'available' P in soil by extraction with sodium bicarbonate solution (Section 1.3).

The dry-ashing plus acid digestion of tea leaves (Section 1.1) and the available phosphorus extraction (Section 1.3) are contrasting procedures in regard to the acceptable precision of the determination. In one the sample is weighed to ± 0.1 mg ($\pm 0.02\%$), whereas in the other the 5 cm^3 scoop used for measuring soil, may introduce a variation of ± 0.2 g ($\pm 4\%$). You may like to think about this. Is the sample measuring procedure appropriate for the purpose of the analysis in each case?

It is worth noting that the tea manufacturers have taken care of grinding the tea leaves. Their products are thoroughly homogenised to ensure that the product quality is uniform. Therefore there should be little variation between tea bags of one type of tea. However, it might be of interest to do a further study to find out if the release of Mn and Zn into the tea infusion is affected by the fineness of grinding.

1.1 Mineralisation of Tea Samples

The inorganic components (total Mn and Zn) that are to be determined in the tea samples must first be brought into solution. Dry-ashing in a muffle furnace destroys the organic matter in the samples leaving a mineral ash. The ash contains the elements in the form of oxides which dissolve in hydrochloric acid leaving an insoluble residue of silica.

Glassware used routinely in the laboratory is easily contaminated by traces of metallic elements which become adsorbed onto the surface of the glass. They are not easily removed by ordinary washing. To remove previous contamination and prevent it from invalidating the results, the glassware has to be washed with nitric acid immediately before it is used for trace metal analysis.

1.1.1 Reagents. Refer to safety data in the Appendix and use appropriate personal protective equipment.

- Nitric acid (approx. 0.5 molar HNO_3).
  Add 32 cm^3 of 69.5% w/w nitric acid (specified for trace metal analysis) to water and dilute to 1 litre.
- Nitric acid (approx. 1.0 molar HNO_3).
  Add 64 cm^3 of 69.5% w/w nitric acid (specified for trace metal analysis) to water and dilute to 1 litre.
- Hydrochloric acid (approx. 1.0 molar HCl).
  Add 85.5 cm^3 of 35–38% w/w hydrochloric acid (specified for trace metal analysis) to water and dilute to 1 litre.
- Hydrochloric acid (approx. 0.1 molar HCl).
  Add 8.6 cm^3 of 35–38% w/w hydrochloric acid (specified for trace metal analysis) to water and dilute to 1 litre.

1.1.2 Equipment Required.

- Muffle furnace with temperature control and able to maintain a temperature of 475 °C.
- Hotplate with temperature control and able to maintain a temperature of 100 °C.
- 30 cm^3 porcelain or silica crucibles (sufficient for the number of samples), individually identified by *heat stable* number markings.
- 50 cm^3 volumetric flasks (sufficient for the number of samples), marked with sequential numbers.
- Filter papers (Whatman No. 2, 11 cm).
- Polypropylene funnels (6.5 cm diameter).

1.1.3 Analytical Procedure.

- Having done the risk assessment, make sure that you wear personal protective equipment. Prepare all the crucibles, volumetric flasks, pipettes and beakers to be used, in advance, by soaking overnight in 0.5 M nitric acid. Then rinse three times with high purity deionised water and dry.
- Make a table in your laboratory notebook of the sequence number, sample numbers and corresponding crucible numbers (Table 5.1). Mark the volumetric flasks with sequence numbers. Include replicate samples, check samples (if available) and blank samples as required. (A blank sample is a crucible and reagents, but with no tea material, taken through the complete procedure.)

- Weigh 0.5 g ($\pm$ 0.1 g) of each sample into a 30 cm^3 porcelain crucible and make a note of the exact weight in your notebook (Table 5.1).
- Ash the samples in a muffle furnace overnight (16 hours) at 475 °C.
- Allow to cool and add, by pipette, 5 cm^3 of 1 M nitric acid (HNO$_3$) solution.
- In a fume-cupboard, evaporate the samples to dryness on a hotplate at 100 °C.
- Return the samples to the muffle furnace at 475 °C for 15–30 minutes.
- Allow to cool and add 10 cm^3 of 1 M hydrochloric acid (HCl).
- Filter each sample solution into a 50 cm^3 volumetric flask using Whatman filter papers, No. 2.
- Wash the crucible and filter paper with several portions of 0.1 M HCl.
- Make the volume in the flask up to 50 cm^3 with 0.1 M HCl; stopper and mix well.
- Store the sample solutions for the subsequent determinations of Mn and Zn.

Table 5.1 *Example of a sample list for digestion of samples*

Sequence no.	Sample lab no.	Crucible no.	Sample weight (g)
1	*e.g.* T2-1	*e.g.* 23	*e.g.* 0.502
2	T2-2	28	0.499
3			
4			
5			
$n-1$	Check sample	54	0.495
n	Blank sample	46	

1.2 Hot Water Extraction of Tea Samples (Infusion)

In this test the tea-making procedure is standardised for the purpose of finding out if different amounts of manganese and zinc are released into the infusion from different types of tea. Tea leaves may contain large amounts of manganese and zinc but only relatively small proportions get into the infusion. Most of the metallic elements remain as insoluble complexes with organic compounds in the leaves. However, tea may still be an important dietary source of manganese and zinc.

A tea bag contains about 3.4 g of tea and a large cup may contain about 250 cm^3 of water. It is not convenient to measure and dispense boiling water, so the normal domestic procedure has been adapted for the laboratory.

1.2.1 Reagents. Refer to safety data in the Appendix and use appropriate personal protective equipment.

- Nitric acid (approx. 0.5 molar HNO$_3$).
  Add 32 cm^3 of 69.5% w/w nitric acid (specified for trace metal analysis) to water and dilute to 1 litre.
- Nitric acid (approx 1.0 molar HNO$_3$).
  Add 64 cm^3 of 69.5% w/w nitric acid (specified for trace metal analysis) to water and dilute to 1 litre.
- Hydrochloric acid (approx. 0.28 molar HCl).
  Add 24 cm^3 of 35–38% w/w hydrochloric acid (specified for trace metal analysis) to water and dilute to 1 litre.

1.2.2 Equipment Required.

- Muffle furnace with temperature control and able to maintain a temperature of 475 °C.
- Hotplate with temperature control and able to maintain a temperature of 100 °C.
- Conical flasks 500 cm^3.
- Conical flasks 100 cm^3.
- 30 cm^3 porcelain or silica crucibles (sufficient for the number of samples), individually identified by *heat stable* number markings.
- 30 cm^3 sample vials with caps.
- Measuring cylinder 250 cm^3.
- Weighing papers or disposable weighing boats.
- Hotplate with temperature control.
- Laboratory clock or timer.
- Plastic tea strainer.
- Polypropylene funnel (6.5 cm diameter).

1.2.3 Analytical Procedure. Note: plan how you will keep to the sequence of actions and maintain the timing of the extraction required in steps (e) to (h).

  a) Prepare all the flasks, crucibles, vials, pipettes, strainer, funnel and beakers to be used, in advance, by soaking overnight in 0.5 M nitric acid. Then rinse them three times with high purity water and dry them.

Table 5.2 *Sample list for analysis (results to be inserted after determinations in Chapter 6)*

Sequence no.	Sample lab no.	$P\ mg\ l^{-1}$	
		in extract	in soil
1	e.g. S4-10	e.g. 2.21	44.2
2	S4-11	1.52	30.4
3			
4			
5			
$n-1$	Check sample	2.55	51.0
n	Blank sample	0.01	0.2

b) Prepare a new table in your laboratory notebook, similar to Table 5.2, allowing for repeat samples, check samples, if available, and blank samples. Mark both sets of conical flasks, volumetric flasks and sample vials with sequence numbers.

c) Weigh 3.00 g (±0.01 g) tea leaves onto labelled weighing papers or weighing boats.

d) Using a measuring cylinder, put 200 cm³ deionised water into a labelled 500 cm³ conical flask.

e) Place the conical flask on the hotplate and bring the water just to the boil.

f) Remove the conical flask from the heat and immediately add the weighed sample of tea to it.

g) Swirl the flask to mix the infusion and allow to stand for 5 minutes ±15 seconds.

h) Swirl the infusion thoroughly again and decant a portion (about 30 cm³) through the tea strainer and polypropylene funnel, into a 100 cm³ conical flask.

i) Allow to cool and pipette 10 cm³ of infusion from the 100 cm³ conical flask into a labelled crucible.

j) In a fume-cupboard, evaporate the samples in the crucibles to dryness on a hotplate at 100 °C.

k) Allow to cool, add by pipette 5 cm³ of 1 M nitric acid (HNO_3) solution and evaporate to dryness again.

l) Place the crucibles in the muffle furnace at 475 °C for 15–30 minutes.

m) Allow to cool and add 10 cm³ of 0.28 M hydrochloric acid (HCl) to dissolve the residue in the crucible.

n) Transfer the dissolved residues to labelled sample vials and keep them for the subsequent determinations.

1.3 Available Phosphorus Extraction from Soil

In order to assess the availability of P in a soil we ought to make an attempt to take into account the concentration of P in the soil solution, the amount of P in the soil minerals, the amount adsorbed on the mineral surfaces and the rate at which it can be released into solution. This, however, is too difficult for a routine procedure.

ADAS, the advisory service for farmers in England and Wales, has adopted a method of P-extraction from soil using a sodium hydrogen carbonate solution. This extracts the P in the soil solution and the P that is easily desorbed from soil particles. By correlation of this chemical extraction method with extensive field trials of fertiliser response on different soil types, the advisory services are able to make fertiliser recommendations for farmers.

The solution used is 0.5 M $NaHCO_3$. It is adjusted to pH 8.5 with NaOH and contains 2.5 mg polyacrylamide per litre. The hydrogen carbonate ion displaces adsorbed phosphate from soil surfaces and also may aid the dissolution of calcium phosphates by maintaining the calcium concentration in solution at a low level because dissolving calcium precipitates as calcium carbonate.

The polyacrylamide is added to flocculate any clay which is dispersed by the presence of sodium and a high pH, so that when the suspension is filtered a clear solution is obtained. Some soils release soluble organic matter during shaking which colours the solution and could interfere with the measurement of phosphate. So, to decolorise the extracts, activated charcoal is added during the extraction procedure.

1.3.1 Reagents. Refer to safety data in the Appendix and use appropriate personal protective equipment.

- 0.5% Polyacrylamide solution.
  Heat about 80 cm^3 of distilled water on the magnetic stirrer/hot-plate. Do not allow to boil. While stirring slowly, sprinkle 0.5 g of polyacrylamide into the beaker. Continue to stir slowly for a few hours. Turn the stirrer and hot-plate off, and leave to stand overnight. Transfer to a 100 cm^3 volumetric flask and make up to the mark with distilled water, mix well. Store the solution in a plastic bottle.
- 0.5 M Sodium hydrogen carbonate pH 8.5.
  For each litre of extracting solution dissolve 42 g of sodium

hydrogen carbonate (grade: specified for analysis) in about 400 cm^3 of distilled water. Add 0.5 cm^3 of polyacrylamide solution and make up nearly to 1 litre. Check the pH of the solution, and adjust to pH 8.5 with 1 M sodium hydroxide. Make up to the mark with distilled water and mix well. This is sufficient to extract 10 samples. When a large quantity of extracting solution is required, make the reagent in a large calibrated polythene aspirator.

- Activated charcoal powder – specified to be *low in P* (for example Darco G60).

1.3.2 Equipment Required.

- Reciprocal shaker, giving equivalent to about 275 strokes of 25 mm per minute.
- Magnetic stirrer/hotplate.
- pH meter and buffer solutions for preparing the extracting solution.
- Numbered 175 cm^3 glass bottles, wide-necked, with caps.
- 100 cm^3 measuring cylinder.
- 5 cm^3 scoop for soil samples.
- 5 cm^3 scoop for charcoal powder.
- Numbered polystyrene or glass sample tubes (30 cm^3).
- Filter papers, No. 2, 11 cm.

1.3.3 Analytical Procedure.

- Prepare in your notebook a page similar to Table 5.2; include repeat samples and check samples, if available, and blank samples.
- Scoop 5 cm^3 of the 2 mm sieved soil into a numbered 175 cm^3 glass bottle.
- Add a 5 cm^3 scoop of activated charcoal powder.
- Add 100 cm^3 of the NaHCO$_3$ extracting solution to each sample using the measuring cylinder. The blank samples have the 5 cm^3 scoop of activated charcoal and 100 cm^3 of extracting solution in a bottle, but no soil.
- Fit a cap and make sure that it is screwed on firmly.
- Place the bottles on the reciprocal shaker. Shake for 30 mins.
- Fold twice as many No. 2, 11 cm filter papers as bottles.
- Arrange the numbered sample tubes for phosphate on the bench and rest two folded filter papers in each tube.
- Filter the extracts into the polystyrene tubes, collecting about 20 cm^3 and discarding the rest. The solutions should be clear, though they may be coloured. Refilter any turbid filtrates.
- Cap the tubes, shake to mix the contents and keep the extracts in the refrigerator until you are ready to do the determination of phosphate in solution (Chapter 6).

CHAPTER 6

Determinations

1 MAKING STANDARD SOLUTIONS

As previously discussed (Chapter 2), much of the work in analytical chemistry involves the manipulation of weights and volume measurements. Frequently, calculations are concerned with concentrations; as in:

- The amount (weight) of a component in a unit weight or a unit volume of the sample.
- The amount of the component in a unit volume of solution.

Most instrumental methods of analysis do not give an absolute measure of the concentration of the component we are determining (*i.e.* the 'analyte'). Usually instruments can give only a comparative measurement. We have to compare the response of the instrument (perhaps a meter reading) to a sample solution with the response it shows to solutions of known concentrations. The solutions of known concentrations that we use to calibrate the response of the instrument are the 'standard solutions' for the determination.

It is important that the standard solutions are made very accurately, or the results of the analysis cannot be accurate. It is also important that the other constituents in the solutions are as similar as possible in all the standard solutions, blank solutions and sample solutions. These other constituents make the solution 'matrix'. If the solution matrices are not similar, the instrument may not respond consistently to different concentrations of the analyte.

Of course it is not possible to make all the solution matrices identical, because we cannot know all that is in the sample solutions. However, if the sample extract or digest solutions are in 1 M HCl, for example, then all the standard solutions, blank solutions and check solutions should also be in 1 M HCl. One advantage of increasing the sensitivity of an instrument or a technique is the possibility of using more dilute solutions so that matrix interferences can be decreased.

1.1 Exercise in Calculating Concentrations

The following questions are concerned with making standard solutions. It is essential to become familiar with and competent at calculations such as these. Mistakes in calculating and in making standard solutions are often the source of errors in analysis. Try to plan the steps you would use to make the solutions required. Avoid using pipettes smaller than 2 cm^3 or larger than 25 cm^3 because in inexperienced hands these can introduce large errors. You will find proposed answers at the end of the chapter.

  a) Calculate how much potassium nitrate (KNO_3) you need to weigh out in order to make 1 litre of a solution containing 2000 mg l^{-1} N in water (*i.e.* 2000 ppm N).
  b) From this solution, how would you make a series of 100 cm^3 standard solutions containing 0, 5, 10, 15, 20 and 25 mg l^{-1} N in 2 M KCl?
  c) Calculate how much potassium dihydrogen orthophosphate (KH_2PO_4) you need to weigh out in order to make 1 litre of a solution containing 2000 mg l^{-1} P in water (*i.e.* 2000 ppm P).
  d) From this solution, how would you make a series of 250 cm^3 standard solutions containing 0, 4, 8, 12, 16 and 20 mg l^{-1} P in 0.5 M NaHCO$_3$?
  e) Calculate how much ammonium sulfate {$(NH_4)_2SO_4$} you need to weigh out in order to make 1 litre of a solution containing 2000 mg l^{-1} N in water (*i.e.* 2000 ppm N).
  f) From this solution, how would you make a series of 200 cm^3 standard solutions containing 0, 10, 20, 30, 40 and 50 mg l^{-1} N in 0.1 M HCl? Hint: conc. HCl is approximately 11.6 M.

1.1.1 General Hints. You need to refer to a table of atomic weights, which you will find in the catalogues of the major chemical suppliers.

Remember that in making dilutions, if the volume you need to take by pipette is V_1, then:

$$V_1 = V_2 \times C_2/C_1$$

Where

V_2 is the final volume (or the volume of the volumetric flask for the diluted solution),
C_2 is the concentration of the solution that you want at the end, and

C_1 is the concentration of the solution that you are starting with.

So, in making dilutions:

$$\text{the volume you need to take} = \frac{\text{the volume you want to make} \times \text{the concentration you want}}{\text{the concentration you start with}}$$

2 DETERMINATIONS BY ATOMIC ABSORPTION

Atoms absorb light at wavelengths specific to each element and this property is used in the analytical technique of atomic absorption spectroscopy. For example, in the determination of zinc, when a solution containing zinc compounds is sprayed into a flame at about $1000\,^{\circ}C$, zinc atoms are produced. These will absorb light at the specific wavelengths associated with zinc.

If the light comes from a source made from zinc it contains a very high proportion of wavelengths that are absorbed by zinc atoms. In atomic absorption spectroscopy the light source used is a hollow cathode lamp, specially made for each element to be determined. Measurement of absorbance of the light from a *zinc* hollow cathode lamp gives a very selective method for the measurement of the concentration of *zinc* in a solution introduced into the flame (Figure 6.1).

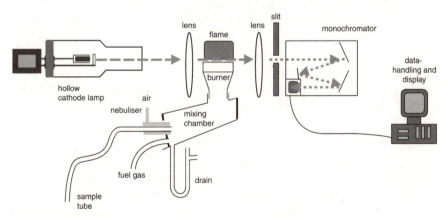

Figure 6.1 *Components of an atomic absorption spectrometer*

However, the instrument has to be calibrated with solutions of known zinc concentration (zinc standard solutions) each time it is used to measure the zinc concentrations in sample solutions. Similarly, the instrument has to be calibrated with manganese standard solutions each time it is used to measure manganese concentrations in sample solutions. In this case the solutions from the ashings and hot water extracts of tea samples are in 0.28 M HCl. So remember, the standard solutions for Zn and Mn need to be made up in 0.28 M HCl as well.

2.1 Determinations of Mn and Zn in Solution

2.1.1 Reagents. Refer to safety data in the Appendix and use appropriate personal protective equipment.

- Hydrochloric acid (approx. 1.4 M HCl).
  Add 120 cm^3 of 35–38% w/w hydrochloric acid (specified for trace metal analysis) to 500 cm^3 water and dilute to 1 litre.
- Hydrochloric acid (approx. 0.28 M HCl).
  Add 24 cm^3 of 35–38% w/w hydrochloric acid (specified for trace metal analysis) to water and dilute to 1 litre.

2.1.2 Equipment Required. Atomic absorption spectrometer (AAS) with hollow cathode lamps for Mn and Zn.

2.1.3 Standard Solutions.

- Stock standard solutions:
  - Manganese stock standard solution: 1000 mg l^{-1} Mn (specified as high purity reagent for AA spectroscopy).
  - Zinc stock standard solution: 1000 mg l^{-1} Zn (specified as high purity reagent for AA spectroscopy).
- Intermediate standard solutions:
  - Manganese intermediate standard solution, 20 mg l^{-1} Mn: pipette 5 cm^3 of stock 1000 mg l^{-1} Mn solution into a 250 cm^3 volumetric flask. Make up to volume with water, stopper and mix well.
  - Zinc intermediate standard solution, 8 mg l^{-1} Zn: pipette 2 cm^3 of stock 1000 mg l^{-1} Mn solution into a 250 cm^3 volumetric flask. Make up to volume with water, stopper and mix well.
- Working standard solutions:
  - Manganese working standard solutions; 0.0, 1.0, 2.0, 3.0, 4.0 and 5.0 mg l^{-1} Mn in 0.28 M HCl: pipette 0, 5, 10, 15, 20 and 25 cm^3 of the 20 mg l^{-1} Mn intermediate standard solution into a set of 100 cm^3 volumetric flasks. Add to each flask 20 cm^3 of 1.4 M HCl, make up to volume with water, stopper and mix well.
  - Zinc working standard solutions; 0.0, 0.4, 0.8, 1.2, 1.6 and 2.0 mg l^{-1} Zn in 0.28 M HCl: pipette 0, 5, 10, 15, 20 and 25 cm^3 of the 8 mg l^{-1} Zn intermediate standard solution into a set of

100 cm³ volumetric flasks. Add to each flask 20 cm³ of 1.4 M HCl, make up to volume with water, stopper and mix well.

2.1.4 Analytical Procedure.

Manganese
- Follow the advice and instructions of the laboratory supervisor in setting-up, optimising and using the atomic absorption spectrometer for determinations of Mn. From the safety data sources available, make a risk assessment of the use of acetylene gas.
- Calibrate the instrument with the Mn working standard solutions and record the concentration of Mn in all the blank, check and sample solutions. After every 10 readings, measure the concentration of Mn in a standard solution and the absorbance of the standard zero. If it is necessary to re-set the zero or re-calibrate the instrument, repeat the previous 10 readings.
- If the concentration of Mn in any sample is greater than 5 mg l⁻¹, a dilution will be needed. By pipette take 5 cm³ of the extract and place it in a labelled 50 cm³ volumetric flask. Make up to 50 cm³ with 0.28 M HCl solution and measure the absorbance as for the other solutions. This solution is a 10-fold dilution of the original and so the measured concentration has to be multiplied by 10.

Zinc
- Set-up and optimise the AAS instrument for determinations of zinc. Calibrate the instrument with the zinc working standard solutions and measure the concentration of zinc in all the blank, check and sample solutions, as for manganese.
- If the concentration of Zn in any sample is greater than 2 mg l⁻¹, a dilution will be needed. By pipette take 5 cm³ of the extract and place it in a labelled 50 cm³ volumetric flask. Make up to 50 cm³ with 0.28 M HCl solution and measure the absorbance as for the other solutions. This solution is a 10-fold dilution of the original and so the measured concentration has to be multiplied by 10.

2.1.5 Calculation and Evaluation. Calculate the concentrations of Mn and Zn in the tea leaves and the infusions and evaluate the results with regard to the purpose of the investigation decided (Chapter 2, Section 3.1).

3 COLORIMETRIC DETERMINATIONS

Colorimetric analysis depends upon the production of a coloured solution in which the depth of colour changes quantitatively as the concentration of analyte changes. We can measure the depth of colour in solution as an indirect measurement of the concentration of analyte (Figure 6.2). In this case the analyte is phosphorus in the form of phosphate.

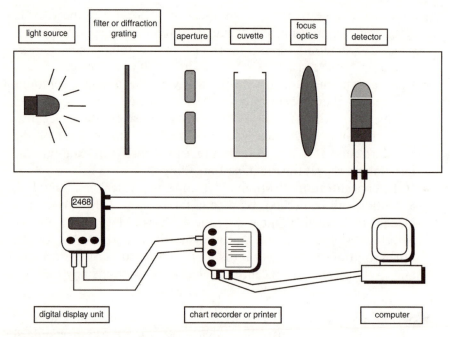

Figure 6.2 *Components of a colorimeter*

3.1 Determination of P in Solution

In this method for the determination of phosphorus in solution, phosphorus (in the form of phosphate) combines with acid ammonium molybdate, forming phosphomolybdic acid. Ascorbic acid reduces the phosphomolybdic acid to give a blue-coloured complex. The antimony salt acts as a catalyst, speeding up the formation of the coloured complex.

We can determine the concentration of phosphorus in a sample solution by comparing the intensity of the blue colour it produces with the intensities produced by a series of solutions containing known amounts of phosphorus. The method can be used for a wide range of applications and extractions.

The main criterion for adapting the method to various extracts is to make the standard solutions and sample solutions up in the *same* matrix.

3.1.1 Reagents. Refer to safety data in the Appendix and use appropriate personal protective equipment.

- Acid molybdate solution.
  With care add 51 cm³ (94 g) sulfuric acid to 500 cm³ water in a conical flask and swirl to mix. Dissolve separately in water 4.3 g l⁻¹ ammonium molybdate and 0.4 g l⁻¹ antimony potassium tartrate. Add the ammonium molybdate to the sulfuric acid and mix, then add the antimony potassium tartrate solution and mix again. Cool the solution, dilute to 1 litre with water and mix well.
- Ascorbic acid solution (1% w/w).
  Dissolve 3 g ascorbic acid in water and make up to 300 cm³ with water, using a measuring cylinder.
- Mixed reagent.
  Mix 200 cm³ molybdate solution with 300 cm³ ascorbic acid solution using a measuring cylinder.

3.1.2 Equipment Required. A colorimeter with a red filter or spectrophotometer set to a wavelength of 660 nm, and 1 cm cuvettes. We choose the filter, or wavelength, so that the blue-coloured solution will absorb a greater proportion of the light passing through it. This increases the 'sensitivity' of the method. We refer to the colour of light that a coloured solution absorbs most strongly as the solution's 'complementary' colour.

All glassware is soaked in phosphate-free detergent (*e.g.* Decon 90) overnight to remove any adsorbed phosphate.

3.1.3 Standard Solutions

- Standard 1000 mg l⁻¹ P.
  Dry potassium dihydrogen orthophosphate (grade: specified for analysis) at 105 °C for 2 hours. Cool in a desiccator. Weigh 4.3938 g into a beaker, dissolve it in deionised water and transfer it to a 1 litre volumetric flask. Make up to the mark with water, stopper and mix well.
- Standard 20 mg l⁻¹ P.
  By pipette, take 10 cm³ of the standard 1000 mg l⁻¹ P solution and put it into a 500 cm³ volumetric flask. Make it up to the mark with water, stopper and mix well.
- Working standard solutions 0–5 mg l⁻¹ P.
  Pipette 0, 5, 10, 15, 20 and 25 cm³ of the standard 20 mg l⁻¹ P solution into 100 cm³ volumetric flasks. To each flask add 4.2 g sodium hydrogen carbonate (NaHCO₃) and add some water to dissolve it. Make each up to the mark, stopper and mix well.

3.1.4 Analytical Procedure. In your notebook draw up a work sheet, similar to that in Table 6.1, including standard solutions, standard checks, blanks and extracts, ending with a final standard check and 'zero check'.

From your standard set of 0 to 5 mg l^{-1} P solutions, take 2 cm³ of each standard solution by syringe or pipette into a set of 30 cm³ glass tubes. From a dispenser or a pipette add 10.0 cm³ of mixed ascorbic acid molybdate reagent. Caution: the acid in the mixed reagent reacts with the sodium hydrogen carbonate in the solutions, causing an effervescence of carbon dioxide (CO_2). Swirl the tube to allow the CO_2 to escape.

Repeat for all the standard solutions, blank solutions and sample solutions. Then swirl all the tubes again.

Secure the caps and shake each tube well. Allow to stand for at least 30 minutes for the development of the blue colour.

Table 6.1 *Sequence table for colorimetric determination of P*

Sequence no.	Solution	Absorbance	Concentration mg l^{-1} P
1	std blank		0
2	std 1 mg l^{-1}		1
3	std 2 mg l^{-1}		2
4	std 3 mg l^{-1}		3
5	std 4 mg l^{-1}		4
6	std 5 mg l^{-1}		5
7	sample 1		
8	sample 2		
9	sample 3		
10	sample 4		
11	sample 5		
12	sample 6		
13	sample 7		
14	sample 8		
15	sample 9		
16	sample 10		
17	std 4 mg l^{-1} check		4
18	'zero check'		–
19	sample 13		
20	sample 14		
etc.			

3.1.5 Measuring the Absorbance.

- Read the absorbance of all the solutions and sample solutions on a colorimeter with a red filter or spectrophotometer set to a wavelength of 660 nm. In using the instrument you must observe the advice and instructions of the laboratory supervisor. Usually, the following method is applicable:
  - You are measuring P concentrations in the range of 0 to 5 mg P per litre and you must be scrupulously clean in all the operations. Always hold the colorimeter cuvettes by the top 1 cm. Fingerprints or solution on the side walls will affect the absorbance readings.
  - Rinse two colorimeter cuvettes with a little of the solution from the standard blank (0 mg l^{-1} P). Fill them to about 3/4 full and dry the side walls with tissue paper.
  - Put one into the colorimeter and set the reading to zero. Measure the absorbance of the other one and keep that one as the 'zero check'.
  - Rinse the first cuvette with the next solution (standard 1.0 mg l^{-1} P), fill it to 3/4 full with the solution, dry it and measure and record the absorbance. Make sure that the cuvette is placed in the holder the same way round each time.
  - Continue to read and record the absorbance of all the standard, blank and sample solutions. After every 10 absorbance readings, measure the absorbance of a standard check and use the 'zero check' cuvette to check the zero setting.
- If a solution from your extracts gives an absorbance reading higher than the 5 mg l^{-1} P standard solution, a dilution of the original extract will be needed (never make a dilution by diluting the coloured solution in the cuvette):
  - By pipette take 5 cm^3 of the original extract, place it in a labelled 50 cm^3 volumetric flask and make it up to 50 cm^3 with $NaHCO_3$ extracting solution;
  - From this solution, take 2 cm^3 for colour development and measure the absorbance as for the other solutions. This solution is a 10-fold dilution of the original and so the measured concentration has to be multiplied by 10.

3.1.6 Calculation and Evaluation.

- Calculate the P content of the sample solutions by making a calibration graph of P concentration (*x*-axis) against absorbance (*y*-axis) for the standard solutions. The graph shows the concentration of P in the extracts, expressed as mg per litre of solution. It should be a straight line or a smooth curve. It should also pass through or very near the zero for both axes. If the line is an irregular curve, then something is most likely wrong with the standard solutions and the graph unacceptable for subsequent calculations.

- Calculate the amount of P in the 100 cm^3 of extract, which came from 5 cm^3 of dry soil, and so calculate the concentration of available P in the soil, expressed as mg P per litre (dm^3) of soil.
- Calculate the concentrations of available P in the soil samples and evaluate the results in regard to the purpose of the investigation decided in Chapter 2. Convert the values into amounts of P per hectare based on the assumption that:
  - 1 litre of dry soil in the lab is 1 kg of dry soil.
  - 1 hectare of soil to plough depth (20 cm) has a volume of:
    $$10^4 \text{ m}^2 \times 0.2 \text{ m} = 2000 \text{ m}^3$$
  - Its bulk density in the field is 1.3 tonnes m^{-3} and 1 tonne = 1000 kg.
- Express the results as kg P per hectare per 20 cm depth.

4 PROPOSED ANSWERS

4.1 Calculation Exercise in Section 1.1

Making standard solutions (avoiding the use of pipettes smaller than 2 cm^3).

4.1.1 Potassium Nitrate Solution (a). From molecular weight of KNO_3, 101.103 g KNO_3 contain 14.0067 g N.

So for 2 g N we need $2 \times 101.103/14.0067$ g KNO_3, *i.e.* 14.436 g KNO_3.

4.1.2 Standard Solutions of N in 2M KCl (b). Overall dilution for the 5 mg l^{-1} solution is 2000/5 *i.e.* 400 $\times$. This is too big a dilution to make in one step, so we need an intermediate dilution – suggest 20 $\times$ dilution. This would give us an intermediate standard solution of 100 mg l^{-1}. Then the volume of this 100 mg l^{-1} solution that we need to make 100 cm^3 of 5 mg l^{-1} is: $100 \times 5/100$ cm^3 *i.e.* 5 cm^3.

So the amounts of the 100 mg l^{-1} solution that we need to make the series of 100 cm^3 of standard solutions containing 0, 5, 10, 15, 20 and 25 mg l^{-1} N are: 0, 5, 10, 15, 20 and 25 cm^3 respectively.

To make each standard solution 2 M with respect to KCl, we need 2×7.455 g KCl per 100 cm^3 *i.e.* 14.91 g KCl in each 100 cm^3 standard solution.

Checking back, we need to have more than 75 cm^3 of our 100 mg l^{-1} intermediate standard solution. With allowance for rinsing the pipettes, 100 cm^3 of it may not be enough, so it would be better to make 200 cm^3 of the intermediate. For this we need: 200 cm^3 × 100/2000 *i.e.* 10 cm^3.

So to make the 100 mg l^{-1} N intermediate standard solution, take 10 cm^3 of the original 2000 mg l^{-1} N solution in a 200 cm^3 volumetric flask, make it up to volume with water, stopper and mix well.

4.1.3 Potassium Dihydrogen Orthophosphate Solution (c). Molecular weight of KH$_2$PO$_4$ is 136.086, atomic weight of P is 30.9738.

So for 2 g of P we need 2 × 136.086/30.9738 g of KH$_2$PO$_4$ *i.e.* 8.7871 g KH$_2$PO$_4$.

4.1.4 Standard Solutions of P in 0.5 M NaHCO$_3$ (d). To make a 4 mg l^{-1} standard solution from a 2000 mg l^{-1} stock solution we need to dilute it by 500 times. Again this is too much for one step. We can make a 10 × dilution, followed by a 50 × dilution; and the 50 × dilution is convenient for the final volume of 250 cm^3.

So, make an intermediate standard of 200 cm^3 of 200 mg l^{-1}: volume needed to take by pipette is 200 cm^3 × 200/2000, *i.e.* 20 cm^3 in a 200 cm^3 flask made up to volume with water.

Then, the volume of this intermediate standard solution that we need to make 250 cm^3 of 4 mg l^{-1} P is: 250 cm^3 × 4/200 *i.e.* 5 cm^3.

So the amounts of the 200 mg l^{-1} solution that we need to make the series of 250 cm^3 of standard solutions containing 0, 4, 8, 12, 16 and 20 mg l^{-1} P are: 0, 5, 10, 15, 20 and 25 cm^3 respectively.

To make each solution 0.5 M with respect to NaHCO$_3$, we need to dissolve 0.5 × 84.0069 × 250/1000 g in each 250 cm^3 standard solution, *i.e.* 10.50 g NaHCO$_3$ in each, before making it up to the mark and mixing.

4.1.5 Ammonium Sulfate Solution (e). Molecular weight of (NH$_4$)$_2$SO$_4$ is 132.134 and the atomic weight of N is 14.0067, so 132.134 g of (NH$_4$)$_2$SO$_4$ contains 28.0134 g N.

For 2 g N we need 132.134 × 2/28.0134 g (NH$_4$)$_2$SO$_4$, *i.e.* 9.4336 g (NH$_4$)$_2$SO$_4$.

4.1.6 Standard Solutions of N in 0.1 M HCl (f). To make a 10 mg l^{-1} solution from a 2000 mg l^{-1} solution we need to dilute it by 200 times. This is too big a dilution to do in one step. We can make a 5 × dilution, followed by a 40 × dilution. The 40 × dilution is convenient for the final volume of 200 cm^3.

So, make an intermediate standard of 200 cm^3 of 400 mg l^{-1} N:

volume needed to take by pipette is 200 cm^3 × 400/2000, *i.e.* 40 cm^3 in a 200 cm^3 flask made up to volume with water.

Then, the volume of this intermediate standard solution that we need to make 200 cm^3 of 10 mg l^{-1} N is: 200 cm^3 × 10/400 *i.e.* 5 cm^3.

So the amounts of the 400 mg l^{-1} N solution that we need to make the series of 200 cm^3 of standard solutions containing 0, 10, 20, 30, 40 and 50 mg l^{-1} are: 0, 5, 10, 15, 20 and 25 cm^3 respectively.

To make each standard solution 0.1 M with respect to HCl, first make 200 cm^3 of 1 M HCl by diluting 17 cm^3 of conc. HCl to 200 cm^3. Then add 20 cm^3 of this 1 M HCl to each 200 cm^3 standard solution, before making it up to the mark with water and mixing.

Literature

1   Eurachem/CITAC Guide, *Quantifying Uncertainty in Analytical Measurement*, 2nd ed., see website *http://www.measurementuncertainty.org/* (or ISBN 0-948-29621-5).

2   Health & Safety Executive (1999), *The Control of Substances Hazardous to Health Regulations (1999)*, Health Directorate, Health & Safety Executive, (SI 437 1999) ISBN 0-11-082087-8.

3   Potter, G. W. H. (1995), *Analysis of Biological Molecules*, Chapman & Hall, ISBN 0-412-49050-1.

4   Prichard, E. (1995), *Quality in the Analytical Chemistry Laboratory*, J. Wiley, ISBN 0471-95541-8 (5).

5   Prichard, E. with MacKay, G. M. and Points, J. (eds.) (1996), *Trace Analysis: A Structured Approach to Obtaining Reliable Results*, The Royal Society of Chemistry, ISBN 0-85404-417-5.

6   Prichard, E. (1998), *Tertiary Education Resource Pack for Analytical Chemistry Quality Assurance*, Version 1.2, produced by Laboratory of the Government Chemist, Teddington.

7   Rafferty, D. (1999) *Getting the Message Across: Key Skills for Scientists*, Royal Society of Chemistry and Glaxo–Wellcome.

8   Rowell, D. L. (1994), *Soil Science Methods & Applications*, Longman Scientific and Technical, ISBN 0-582-087848.

9   UKAS (1997), *The Expression of Uncertainty and Confidence in Measurement*, NAMAS publication M 3003, United Kingdom Accreditation Service, Feltham, UK.

10 UKAS (2000), *The Expression of Uncertainty in Testing*, publication Lab 12, United Kingdom Accreditation Service, Feltham, UK.

11 US Department of Energy, Office of Environmental Management (1997), *The Simulated Site Interactive Training Environment (SimSITE)*, *http://etd.pnl.gov:2080/DQO/simsite/home.htm.*

12 Vogel, A. I. (1989), *Vogel's Textbook of Quantitative Chemical Analysis*, 5th ed., revised by Jeffery, G. H., Bassett, J., Mendham, J. and Denny, R. C., Longman Scientific and Technical, ISBN 0-582-44593-7.

13   Weyhe, S. (1997), *Weighing Technology in the Laboratory*, [Sartorius, transl. Zoeanne Poore-Dielschneider], Verlag Moderne Industrie, ISBN 3-478-93168-1.

Appendix: Safety Data for Reagents

*Note: OES is occupational exposure standard and OEL is occupational
exposure limit (based on HSE Guidance note EH40)*

Tox: = Toxicity   Pro: = Protection   Haz: Other known hazards

Acetone
Risks: Can cause serious damage if splashed in eyes. Degreases skin,
possibly causing dermatitis. Vapour narcotic in high concentrations.
Haz: Extremely flammable. Reacts violently with chloroform and
bromoform in the presence of alkalis or in contact with alkaline
surfaces. Decomposes violently in contact with nitric/sulfuric
acid mixtures. Can react violently with oxidising agents.
Tox: LD50 5800 mg kg^{-1} oral (rat). *OES* 2400 mg m^{-3}.
Pro: Safety glasses. Nitrile gloves. Fume-cupboard.

Activated Charcoal Powder
Risks: Irritating to eyes and if inhaled as dust. May irritate skin.
Haz: Can ignite or explode in contact with strong oxidising agents.
Finely divided materials can cause dust explosions. Can react
violently with potassium and sodium metal. Mixtures with
unsaturated oils can explode.
Tox: No data.
Pro: Rubber or plastic gloves. Good laboratory practice.

Ammonium Metavanadate
Risks: Very toxic by ingestion, inhalation and skin contact. Extremely
irritating to eyes and respiratory system.
Haz: No data.
Tox: May cause adverse *mutagenic or teratogenic* effects. *LD50* 160
mg kg^{-1} oral (rat).
Pro: Rubber or plastic gloves. Handle in a fume-cupboard.

Ammonium Molybdate
Risks: May be harmful by ingestion. Irritation to eyes and to respira-
tory system if inhaled as dust.
Haz: No data.
Tox: Evidence of *mutagenic* effects. *OES* Mo 5 mg m^{-3}.
Pro: Rubber or plastic gloves.

Amylase
Risk: May cause allergic reactions in sensitive individuals. Harmful if
ingested in quantity. May irritate eyes.
Haz: No significant hazard.
Tox: Low toxicity. *LD50* 17000 mg kg^{-1} oral (rat).
Pro: Plastic gloves.

Antimony Potassium Tartrate (Antimony Potassium Oxide (+)-Tartrate)
Risks: Harmful by inhalation and swallowing. Irritates the skin and
risk of sensitisation. *Do not eat, drink or smoke while handling
this material* – wash hands and face after handling.
Haz: Suspected *carcinogen*. May give toxic fumes in fire.
Tox: *LD50* 115 mg kg^{-1} oral (rat). *OES* Sb 0.5 mg m^{-3} (long term).
Pro: Safety glasses. Rubber or plastic gloves. Fume-cupboard.

Ascorbic Acid
Risks: Ingestion of large quantities may cause nausea, vomiting and
diarrhoea. Irritating to eyes.
Haz: No data.
Tox: May cause adverse *mutagenic or teratogenic* effects. *LÐ50* 11900
mg kg^{-1} oral (rat).
Pro: Good laboratory practice.

Cleaning Mixture – Chromium Trioxide in 85% w/v Sulfuric Acid
Risks: Causes severe burns to eyes and skin. If ingested causes severe
internal irritation and damage – nausea, vomiting and diarrhoea. Dilute mixture irritates the eyes and skin and may cause
burns and dermatitis. *Corrosive.*
Haz: Never add water to this product. Reacts violently with many
compounds, including nitrates, permanganates, easily oxidised
substances, organic solvents, organic nitro compounds and
peroxides. No evidence of carcinogenic, mutagenic or teratogenic properties.
Tox: *LD50* 2140 mg kg^{-1} oral (rat) (using 25% solution). *OES*
sulfuric acid 1 mg m^{-3}.

Pro: Goggles or face shield. Nitrile gloves. Plastic apron. Sleeves. Fume-cupboard.

Diaminoethanetetraacetic Acid, Disodium Salt
Risk: Harmful if ingested in quantity. May be irritating to eyes.
Haz: No data.
Tox: *LD50* 2000 mg kg^{-1} oral (rat). Exposure limits not assigned.
Pro: Rubber or plastic gloves.

Disodium Hydrogen Orthophosphate, Anhydrous
Risk: May irritate eyes and respiratory system if inhaled as dust. Ingestion of large amounts of phosphate can cause serious disturbances in calcium metabolism.
Haz: No data.
Tox: *LD50* 17000 mg kg^{-1} oral (rat).
Pro: Rubber or plastic gloves.

Disodium Tetraborate
Risks: Harmful if ingested in quantity. Can cause nausea, agitation and spasms. May be irritating to eyes.
Haz: No evidence of carcinogenic, mutagenic or teratogenic effects.
Tox: *LD50* 2260 mg kg^{-1} oral (rat). *OES* disodium tetraborate decahydrate 5 mg m^{-3}.
Pro: Safety glasses. Rubber or plastic gloves. Good laboratory practice.

Hydrochloric Acid
Risks: Skin: severe burns. Eyes: vapour irritates, liquid burns eyes severely. Respiratory tract: severe irritation. Ingested liquid: severe irritation and damage.
Haz: Causes burns. Can explode on addition to solid potassium permanganate. Reacts violently with sodium metal.
Tox: *LD50* 900 mg kg^{-1} oral (rat). *OEL* 5 ppm.
Pro: Safety glasses. Gloves. Fume-cupboard.

Manganese Standard Solution (1000 ppm)
Risks: Contains 3% nitric acid: harmful if ingested in quantity, irritating to skin and eyes. May cause burns if contact is prolonged.
Haz: May give toxic fumes in fire. No evidence of carcinogenic, mutagenic or teratogenic effects.
Tox: *OES* nitric acid 5 mg m^{-3} (long term).
Pro: Safety glasses. Rubber or plastic gloves.

Nitric Acid

Risks: Corrosive. Causes severe burns to eyes (risk of blindness) and skin. If ingested causes severe internal irritation and damage (risk of death). Irritating harmful vapour (risk of oedemas).

Haz: Reacts violently with a large number of compounds and elements. Substances to be avoided: organic combustible substances, oxidisable substances, organic solvents, alcohols, ketones, aldehydes, amines, anilines, nitriles, organic nitro compounds, hydrazine and derivatives, acetylidene, metals, metallic oxides, alkali metals, alkaline earth metals, ammonia, bases, acids, hydrides, halogens, halogen compounds, non-metallic oxides and halides, non-metallic hydrogen compounds, non-metals, phosphides, nitrides, lithium silicide, hydrogen peroxide.

Tox: *OEL* 2 ppm. *OES* 5 mg m^{-3}.

Pro: Safety glasses. Gloves. Fume-cupboard.

Polyacrylamide

Risks: May be harmful if ingested in quantity. May irritate eyes. A permitted food additive.

Haz: Combustible. No evidence of carcinogenic, mutagenic or teratogenic properties.

Tox: No data.

Pro: Safety glasses. Rubber or plastic gloves.

Potassium Dihydrogen Orthophosphate

Risks: May irritate eyes and respiratory system if inhaled as dust. Ingestion of large amounts of phosphate can cause serious disturbances in calcium metabolism.

Haz: No data.

Tox: No data.

Pro: Good laboratory practice.

Sodium Dichromate

Risks: Harmful by ingestion, inhalation and skin contact. Corrosive to skin and eyes. Frequent exposure can cause ulceration, liver and kidney disease and cancer. *Do not eat drink or smoke while handling this material* – wash hands and face after handling.

Haz: May ignite combustible material.

Tox: Chromium(IV) is highly toxic. *Suspected carcinogen. Evidence of mutagenic effects. OES* Cr 0.05 mg m^{-3}. Lethal dose (man) 0.5 g Cr.

Pro: Safety glasses. Rubber or plastic gloves.

Sodium Dodecyl Sulfate (Sodium Lauryl Sulfate)

Risk: Harmful by ingestion. Irritating to eyes and respiratory system if inhaled as dust – causes sneezing (sternutatory). *Weigh and transfer this chemical in fume-cupboard.*

Haz: No data.

Tox: $LD50$ 1288 mg kg^{-1} oral (rat).

Pro: Safety glasses. Rubber or plastic gloves. Fume-cupboard.

Sodium Hydrogen Carbonate

Risks: Ingestion of large quantities may cause nausea.

Haz: No evidence of other hazardous properties.

Tox: $LD50$ 4220 mg kg^{-1} oral (rat).

Pro: Safety glasses. Good laboratory practice.

Sodium Hydroxide

Risks: Corrosive to body tissue, causing burns and deep ulceration. Irritant and harmful as dust. If ingested causes severe internal irritation and damage.

Haz: Dissolves with evolution of heat. Can react vigorously with chloroform/methanol mixtures, strong acids and zirconium. Can cause zinc dust to ignite.

Tox: OES 2 mg m^{-3}.

Pro: Safety glasses. Rubber or plastic gloves. Handle hot solutions in fume-cupboard.

Sulfuric Acid

Risk: Causes severe burns to eyes and skin. If ingested causes severe internal irritation and damage. Dilute acid irritates the eyes and skin and may cause burns and dermatitis. *Corrosive.*

Haz: Can cause acetonitrile to undergo a violent reaction. Can react vigorously or violently with organic nitro compounds and with potassium permanganate, metal halogenates, perchlorates and alkali metals.

Tox: $LD50$ 2140 mg kg^{-1} oral (rat). $LC50$ 0.51 mg l^{-1} inhaled (rat). OEL 1 mg m^{-3}.

Pro: Safety glasses. Nitrile gloves. Fume-cupboard.

Triethylene Glycol (Trigol)

Risk: May be harmful if ingested in quantity. May be irritating to the eyes.

Haz: May react with oxidising materials. No evidence of other hazardous properties.

Tox: *LD50* 17000 mg kg^{-1} oral (rat).
Pro: Safety glasses. Rubber or plastic gloves.

Zinc Standard Solution (1000 ppm)

Risks: Contains 3% nitric acid: harmful if ingested in quantity, irritating to skin and eyes. May cause burns if contact is prolonged.
Haz: May give toxic fumes in fire. No evidence of carcinogenic, mutagenic or teratogenic effects.
Tox: *OES* nitric acid 5 mg m^{-3} (long term).
Pro: Safety glasses. Rubber or plastic gloves.

Subject Index